Springer Theses

Recognizing Outstanding Ph.D. Research

Aims and Scope

The series "Springer Theses" brings together a selection of the very best Ph.D. theses from around the world and across the physical sciences. Nominated and endorsed by two recognized specialists, each published volume has been selected for its scientific excellence and the high impact of its contents for the pertinent field of research. For greater accessibility to non-specialists, the published versions include an extended introduction, as well as a foreword by the student's supervisor explaining the special relevance of the work for the field. As a whole, the series will provide a valuable resource both for newcomers to the research fields described, and for other scientists seeking detailed background information on special questions. Finally, it provides an accredited documentation of the valuable contributions made by today's younger generation of scientists.

Theses are accepted into the series by invited nomination only and must fulfill all of the following criteria

- They must be written in good English.
- The topic should fall within the confines of Chemistry, Physics, Earth Sciences, Engineering and related interdisciplinary fields such as Materials, Nanoscience, Chemical Engineering, Complex Systems and Biophysics.
- The work reported in the thesis must represent a significant scientific advance.
- If the thesis includes previously published material, permission to reproduce this must be gained from the respective copyright holder.
- They must have been examined and passed during the 12 months prior to nomination.
- Each thesis should include a foreword by the supervisor outlining the significance of its content.
- The theses should have a clearly defined structure including an introduction accessible to scientists not expert in that particular field.

More information about this series at http://www.springer.com/series/8790

Martin Klose

Landslide Databases as Tools for Integrated Assessment of Landslide Risk

Doctoral Thesis accepted by the University of Vechta, Germany

Springer

Author
Dr. Martin Klose
Applied Physical Geography
University of Vechta
Vechta
Germany

Supervisor
Prof. Bodo Damm
Applied Physical Geography
University of Vechta
Vechta
Germany

ISSN 2190-5053 ISSN 2190-5061 (electronic)
Springer Theses
ISBN 978-3-319-20402-4 ISBN 978-3-319-20403-1 (eBook)
DOI 10.1007/978-3-319-20403-1

Library of Congress Control Number: 2015942627

Springer Cham Heidelberg New York Dordrecht London

Printed on acid-free paper

Springer International Publishing AG Switzerland is part of Springer Science+Business Media
(www.springer.com)

Parts of this thesis have been published in the following journal articles:

Damm, B., Klose, M., 2014. Landslide Database for the Federal Republic of Germany: A Tool for Analysis of Mass Movement Processes and Impacts. In: Sassa, K., Canuti, P., Yin, Y. (Eds.), Landslide Science for a Safer Geoenvironment, Volume 2: Methods of Landslide Studies. Springer, Berlin, pp. 787–792.

Klose, M., Damm, B., Gerold, G., 2012. Analysis of Landslide Activity and Soil Moisture in Hillslope Sediments using a Landslide Database and a Soil Water Balance Model. GEO-ÖKO 33(3-4), 204–231.

Klose, M., Damm, B., Terhorst, B., 2015. Landslide Cost Modeling for Transportation Infrastructures: A Methodological Approach. Landslides 12, 321–334.

Klose, M., Gruber, D., Damm, B., Gerold, G., 2014. Landslide Susceptibility Modeling on Regional Scales: The Case of Lower Saxony, NW Germany. In: Sassa, K., Canuti, P., Yin, Y. (Eds.), Landslide Science for a Safer Geoenvironment, Volume 2: Methods of Landslide Studies. Springer, Berlin, pp. 437–442.

Klose, M., Gruber, D., Damm, B., Gerold, G., 2014. Spatial Databases and GIS as Tools for Regional Landslide Susceptibility Modeling. Zeitschrift für Geomorphologie NF 58(1), 1–36.

Klose, M., Highland, L., Damm, B., Terhorst, B., 2014. Estimation of Direct Landslide Costs in Industrialized Countries: Challenges, Concepts, and Case Study. In: Sassa, K., Canuti, P., Yin, Y. (Eds.), Landslide Science for a Safer Geoenvironment, Volume 2: Methods of Landslide Studies. Springer, Berlin, pp. 661–667.

Risk varies inversely with knowledge

Irving Fisher, *The Theory of Interest*, 1930

Supervisor's Foreword

Landslide risk is a pressing societal issue that is still poorly understood. A major challenge of risk assessment originates from the difficulty of quantifying risk considering the wide range of landslide types and processes and the various cost factors independent of size or magnitude. Recent studies stress the importance of integrated approaches that use damage statistics and data on societal risk acceptance to explore landslide risk in all its facets. A key to these new approaches are landslide databases that store geospatial and impact-related information on past and current landslides. The availability of Geographic Information Systems (GIS) in recent years has made landslide databases an important tool for spatial inventory and hazard mapping. The full scientific potentials of databases in risk assessment, however, go far beyond the scope of GIS applications, but are still widely underestimated. This relates to a lack of approaches capable of searching database contents for damage or cost information and to derive risk by the systematic fusion of complex data sets from multiple sources. The development of innovative tools for knowledge discovery in landslide databases is critical for assessing landslide risk in integrated perspective.

This doctoral thesis written by Martin Klose is a pioneering research work that makes an excellent contribution to fundamental understanding of landslide risk. The study introduces an analytical framework for integrated risk assessment and new approaches to data integration, modeling, and visualization tailored for use with data sets extracted from landslide databases. "From physical process to economic cost" is the principle of method development in this research work, with the goal of bridging the gap between the analysis of landslide hazard and impact. A key role is played by a landslide susceptibility model that enables to identify and delineate areas at risk of landslides and to assess infrastructure exposure. Temporal landslide hazard is derived from landslide frequency statistics and a hydrological simulation approach to estimate triggering thresholds. These methods are integrated into a powerful toolset for cost survey and modeling that uses historical data to compile, model, and extrapolate damage costs on different spatial scales over time. The combination of this toolset with techniques to analyze fiscal cost impacts supports integrated risk assessment by exploring the economic relevance of landslide losses.

Martin Klose presents in his doctoral thesis a novel approach to landslide risk assessment that constitutes a major scientific advance in a research field critical to global society. The thesis is a brilliant example of cross-cutting and societally relevant Ph.D. research in the Earth Sciences and neighboring disciplines. It is to expect that the thesis will make a global impact, which is already reflected by the attention paid to the journal articles accompanying this research work. Martin Klose has written the thesis with the experience of a four-month research visit at the U.S. Geological Survey in Golden (CO), USA. The cooperative research he made at this world-leading research institute was funded by a scholarship of the German Academic Exchange Service. Martin Klose initiated this partnership with colleagues from the Landslide Hazards Program and is actively participating in global scientific exchange and the consulting of decision makers. The results of this cooperative research and further projects found their way into his doctoral thesis and provide an international perspective on landslide risk. This makes the present thesis a top-level research work of high scientific excellence. It is therefore a great pleasure to nominate Martin Klose for a Springer Thesis Prize.

Vechta
April 2015

Prof. Bodo Damm

Acknowledgments

This doctoral thesis in Applied Physical Geography was prepared at the University of Vechta under the supervision of Professor Bodo Damm. I am grateful to the scientific advice he gave to me in all stages of the Ph.D. research. His mentoring over the past 3 years together with the responsibility he delegated to me has been a great personal benefit. Thanks are also due to Professor Birgit Terhorst who co-supervised the thesis and without her help this research would have never been possible. Professor Jörg Grunert is thanked for his scientific support and for acting as external examiner. I would also like to thank Lynn Highland for her great hospitality during a research visit at the USGS Geologic Hazards Science Center and her valuable assistance in scientific writing and communication.

The doctoral thesis was an essential part of the research project *Identification and Spatial Modeling of Hazard Potentials from Landslides in Lower Saxony* (MWK Niedersachsen 11.2-76202-10-1/07). Much of this Ph.D. research has also been conducted in the framework of the project *Landslide Database for the Federal Republic of Germany* funded by the German Research Foundation (DFG DA 452/6-1). A further part of the research work benefited from a scholarship given by the German Academic Exchange Service for studying in the USA (DAAD D/12/45096). Travel grants were also provided by the Universitätsbund Göttingen e.V. and the University of Vechta. The financial support and research funding of these institutions is gratefully acknowledged.

This Ph.D. research benefited from the scientific and technical assistance from Bachelor and Master students at the Universities of Vechta and Göttingen. I owe special thanks to Daniel Gruber and Philipp Maurischat for their support in conducting the studies. For the pleasant collaboration during my time at the University of Göttingen, I would like to thank Professor Gerhard Gerold, Dorothea Röckl, and Ingrid Kohlrautz-Weinrich. Thanks are also due to Susanne Döhler, Claudia Dornieden, Annette Janßen, and Monika Kolbeck who were great colleagues at the University of Vechta. This doctoral thesis would have never been possible without the support of my parents and brother to whom I am particularly grateful.

Contents

1 Introduction . . . 1
1.1 Landslides—Why So Complex Phenomena? . . . 1
1.2 Overview of Global Landslide Impact . . . 5
1.3 Research Gap . . . 12
1.4 Objectives and Outline of the Study . . . 15
References . . . 19

2 Landslide Databases—State of Research and the Case of Germany . . . 25
2.1 Evolution of Landslide Databases—An Overview . . . 25
2.2 Landslide Database for the Federal Republic of Germany . . . 29
2.2.1 Background and Goals of the Database . . . 29
2.2.2 Structure, Content, and Information Sources of the Database . . . 30
2.2.3 Examples of Regional Database Application . . . 35
References . . . 40

3 Study Area . . . 45
3.1 Regional Setting . . . 45
3.2 Landslide Types, Processes, and Materials . . . 47
References . . . 55

4 Methodology . . . 59
4.1 Soil Water Balance Model . . . 59
4.1.1 Introduction . . . 59
4.1.2 Model Description . . . 61
4.2 Landslide Susceptibility Model . . . 67
4.2.1 Introduction . . . 67
4.2.2 Model Description . . . 71

4.3 Landslide Cost Assessment Model 79
4.3.1 Introduction 79
4.3.2 Model Description 83
References 91

5 Results 107
5.1 Spatial Hazard—Where Do Landslides Occur? 107
5.1.1 Landslide Characteristics and Geofactor Weights 107
5.1.2 Landslide Susceptibility Map 110
5.1.3 Model Assessment 114
5.1.4 Discussion 116
5.2 Temporal Hazard—When and Why Do Landslides Occur? 122
5.2.1 Stability Criteria and Strength Properties of Hillslope Sediments 122
5.2.2 Temporal Development of Soil Water Balance 123
5.2.3 Correlation Between Landslide Activity and Simulated Soil Moisture 124
5.2.4 Critical Soil Moisture Threshold 124
5.2.5 Recurrence Frequency 125
5.2.6 Landslide Volume and Soil Moisture Level 126
5.2.7 Period of Saturation and Initiation Time 127
5.2.8 Discussion 128
5.3 Hazard Impact—How Much Do Landslides Cost? 129
5.3.1 Landslide Losses for Highways in the Upper Weser Area 129
5.3.2 Landslide Losses for Highways in the Lower Saxon Uplands 132
5.3.3 Discussion 134
References 138

6 Synthesis—Towards Integrated Assessment of Landslide Risk 143
6.1 Created Risk—The Role of Human Activity 143
6.2 Are Landslides Economically Relevant? 148
6.3 Conclusions 151
References 154

Chapter 1
Introduction

1.1 Landslides—Why So Complex Phenomena?

Landslides are among the world's most frequent geohazards and pose serious risks to human activity on slopes across the globe (e.g., Brabb 1991; Dilley et al. 2005; Nadim et al. 2006; Hong et al. 2007; Kirschbaum et al. 2010; Petley 2012). As simply defined, a landslide is "the movement of a mass of rock, earth or debris down a slope" (Cruden 1991), with gravity and often water as well being the major driving factors of this geomorphic process (e.g., Sidle and Ochiai 2006; Lu and Godt 2013). The diversity of landslide types is large, ranging from slides in a strict sense to types of movement such as flows, falls, topples, spreads, and complex landslides, a combination of at least two of these movement types (cf. Varnes 1958, 1978; Nemčok et al. 1972; Cruden and Varnes 1996; Dikau et al. 1996; Hungr et al. 2014). Each landslide represents a specific state of activity in a much broader concept of slope stability. The stability of slopes is usually understood as a "physical system that develops in time through several stages", including besides the landslide itself (slope failure), complex pre- and post-failure process mechanisms (Hungr et al. 2014; see also Terzaghi 1950; Vaunat et al. 1994; D'Elia et al. 1998). Various states of landslide activity (active, dormant, reactivated, etc.) can be differentiated (e.g., WP/WLI 1993), and within the broad stability spectrum of slopes, the shift from stable to unstable conditions over time is controlled by predisposition, preparatory, and triggering factors (cf. Crozier 1986; Glade and Crozier 2005a).

The causes and triggers of landslides are diverse (e.g., Wieczorek 1996), varying between the different regions of the world, but with intense or prolonged rainfall (e.g., Guzzetti et al. 2008; Kirschbaum et al. 2012), earthquake shaking (e.g., Keefer 2002; Ugai et al. 2013), and human activity (e.g., Sidle et al. 1985; Nadim et al. 2011) being globally the most widespread causative factors. Rapid urbanization of the world's hillsides today increasingly involves settlement in areas susceptible to landslides while often intensifying landslide susceptibility by slope disturbance itself (cf. Alexander 1989; Pike et al. 2003; Schuster and Highland 2007). Both their close dependency on human activity and the variety of their types and processes make landslides an everyday hazard in many areas worldwide

M. Klose, *Landslide Databases as Tools for Integrated Assessment of Landslide Risk*, Springer Theses, DOI 10.1007/978-3-319-20403-1_1

(Klose et al. 2014a). This distinguishes landslides from related geohazards (earthquakes, storm events, etc.) of which they are a frequent secondary effect (e.g., Harp et al. 2009; Marano et al. 2010), with their losses increasing that of the triggering event significantly (cf. Budimir et al. 2014).

Landslides more than any other geohazard are characterized by a complex distribution in space and time (Fig. 1.1). Each year thousands of landslides occur worldwide, not only in high or low mountain areas (e.g., Korup 2012), but also in parts of the world with little topographic relief (cf. Brabb and Harrod 1989), including the shorelines of oceans (e.g., Lee and Clark 2002; Iadanza et al. 2009), artificial landscapes in lowland areas (e.g., Wichter 2007), and even continental shelves undersea (e.g., Hampton et al. 1996; Masson et al. 2006). The diversity of distribution areas is only one aspect in the complexity of landslide risk; more important, however, are the spatiotemporal patterns in landslide occurrence, especially at local and regional level. Five different spatial and/or temporal patterns of landslide activity are generally identifiable: (i) event-based clustering (Fig. 1.1; see also Cardinali et al. 2000), (ii) seasonal clustering (e.g., monsoon cycle; Petley et al. 2007), (iii) geofactor-oriented clustering (relief, lithology, etc.; cf. Sect. 5.1), (iv) continuous (or episodic) landsliding (slope creep over broad areas; e.g., Hilley et al. 2004), and (v) land use-related distribution (dispersed or clustered; see also Fig. 1.1). Although not with a strict focus to risk analysis, spatiotemporal patterns in landslide occurrence have already been described in related studies as well, including, amongst others, Witt et al. (2010), Rossi et al. (2010), and Tonini et al. (2013).

A unique feature of landslides is their large spectrum of sizes, velocities, and lifetimes (e.g., Malamud et al. 2004; Guthrie and Evans 2007; Crozier 2010). From a global perspective, the size spectrum of a single landslide spans at least nine (areal extent) to more than twelve (volume) orders of magnitude (Guzzetti 2005; Guzzetti et al. 2012). Landslides in their extremes are thus either discrete points in space or large regional phenomena whose deposits cover tens or hundreds of square kilometers (cf. Glade and Crozier 2005b). Alternatively, the velocity of landslides ranges from slow (mm/year) to extremely rapid (m/s) movement, whereby velocity and distribution of activity is often varying within a single landslide (e.g., Cruden and Varnes 1996). Landslides result in landforms that show long persistence in the geomorphic landscape (cf. Guthrie and Evans 2007), with ages of landslide features and deposits reaching up to thousands of years in many cases (e.g., González Díez et al. 1996; Terhorst 2001). The long lifetime of landslides together with their nature to create rough and unstable terrain often makes areas affected by landslides inhabitable for decades or even centuries (e.g., Burke et al. 2002; Van Den Eeckhaut et al. 2010).

Landslide impact in physical terms involves damage to people or property located on a slide mass or in its pathway by burial, collision, and displacement (direct impact). Besides these on-site impacts, there are also impacts experienced off-site (indirect impact), including damage from landslide-induced secondary hazards. The impact of a landslide is either temporally coinciding with its occurrence (instant impact) or emerges in its aftermath (delayed impact) (cf. Glade and Crozier 2005a; Crozier et al. 2013). The degree of landslide impact (landslide intensity) is highly variable and largely depends on the type of landslide, its magnitude, and the vulnerability

Fig. 1.1 Examples of landslide triggering events with a complex distribution of landslides in space and time. **a** Widespread landslide activity on highly developed slopes in southern California (USA) as result of severe winter storms in January and February 2005 (*Photo* J. Godt, USGS). **b** Land use as key factor for causing thousands of landslides during the February 2004 rainstorm on southern North Island, New Zealand (*Photo* G. Hancox, GNS Science)

of the element at risk (building, road, etc.) (e.g., Alexander 1986; Flageollet 1999; Pitilakis et al. 2011). Landslide intensity as "the destructive power of a landslide" (Corominas et al. 2014) is generally difficult to define due to unique problems in parameterization, measurement, and scaling of landslide magnitude (cf. Guzzetti 2005). This is mainly because landslides show significant impact as both: fast-moving

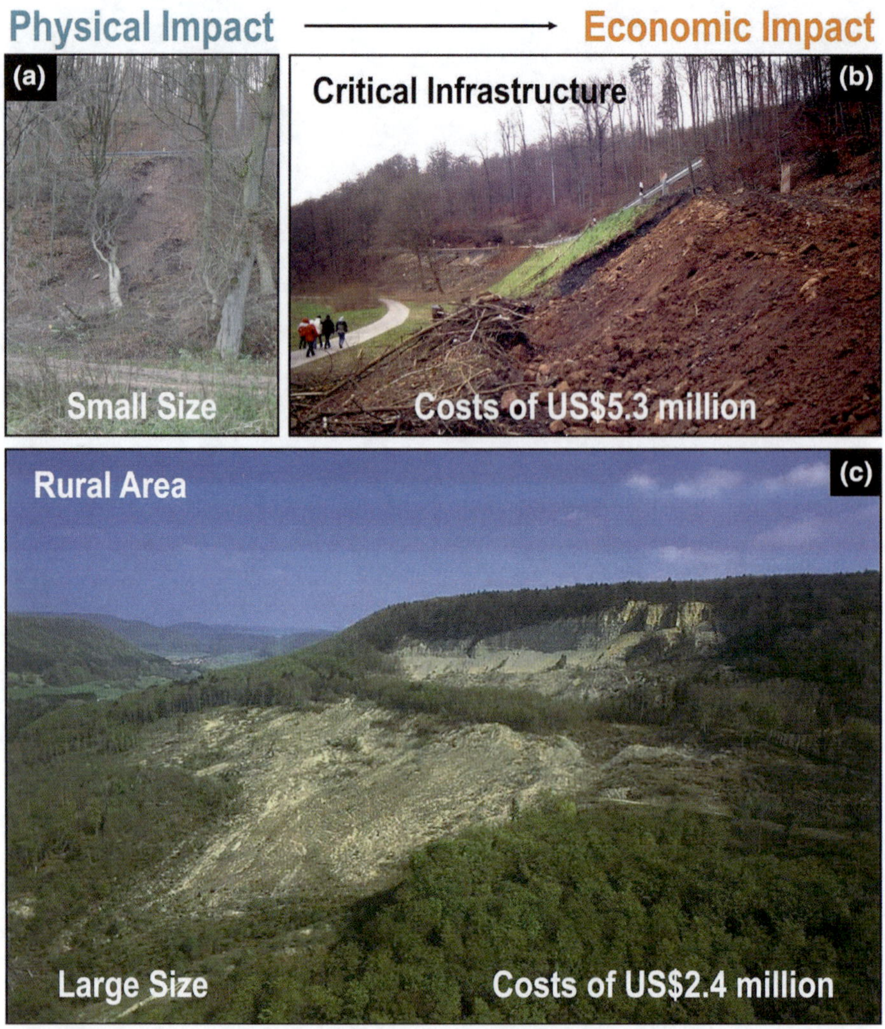

Fig. 1.2 Illustration of the complex relationship between physical landslide processes and the economic impact of landslides. As the examples from NW and SW Germany show, the costs of landslide damage are sometimes independent from landslide magnitude, while cost factors such as landslide location and type of affected infrastructure are often playing an important role: **a**, **b** Landslide damage and repair of highway B 80 or B 3 north of the city of Hann. Münden, Lower Saxony, after repeated landslide damage in the mid-2000s (*Photo* M. Klose; Database B. Damm); **c** the 1983 Mössingen landslide at the Swabian Jurassic escarpment, Baden-Wuerttemberg (*Photo* A. Dieter; cf. Munich Re 1999)

landslides (high magnitude) with large associated damage, and slow-moving landslides (low magnitude) that result in large damage over time as well (e.g., Cruden and Varnes 1996; Urciuoli and Picarelli 2008; Mansour et al. 2011; Antronico et al. 2014).

The relationship between physical landslide impact and economic costs depends on a variety of factors often independent from landslide magnitude (cf. Klose et al. 2014a, b). Most types of landslides result in specific kinds of damage whose translation into monetary losses is beyond simple expression through linear magnitude–damage–cost relationships. A characteristic feature of landslide impact is that there is no strict rule that the larger landslide magnitude, the higher landslide costs (Fig. 1.2; see also Sect. 5.3.1). Thus, even shallow soil slides or small rockfalls affecting highways may result in large costs, while the losses of major landslides in remote rural areas are not necessarily large. Severity of economic impact (direct or indirect) first relates to landslide location (urban or rural) and the question whether critical infrastructure is affected or not (Fig. 1.2; e.g., Blaschke et al. 2000; Geertsema et al. 2009). A second main driver of landslide costs, as case studies indicate (e.g., Cornforth 2005; Hearn et al. 2011; Highland 2012), are the types and methods of post-disaster mitigation, whereas a correlation between direct costs and the damage to or the value of elements at risk is often hard to find (cf. Sect. 5.3.1). These cost factors related to landslide repair and prevention are partly controlled by the level of public and individual risk acceptance and thus the underlying societal conditions (e.g., Fell 1994; Finlay and Fell 1997; Bell et al. 2006; Winter and Bromhead 2012). As a result of disparities in technical and adaptive standards, coping with landslides differs throughout the world. This also causes their impacts and costs to vary geographically, specifically as a function of the region`s level of economic development (cf. Klose et al. 2014a).

1.2 Overview of Global Landslide Impact

Statistics on the death toll and economic losses of landslides are rare to find for above reasons, but those few that are available at national or continental scale clearly illustrate the global significance of landslide impact (Table 1.1; see also Alcántara-Ayala 2014). According to a study from Dilley et al. (2005), an area corresponding to 2.5 % of the world's land surface is prone to landslides. The study states further that 300 million people (5 % of world population) across the globe are living in areas exposed to significant landslide risk. Furthermore, Petley (2012) has found that between 2004 and 2010 landslides claimed more than 30,000 lives, with a strong concentration of landslide fatalities in E- and SE-Asia, the Himalayas, and Central America (Fig. 1.3). These regional clusters of fatal landslides are often considered as global landslide hotspots (Nadim et al. 2006). By referring to data from the Centre of Research for the Epidemiology of Disasters (CRED), Kjekstad and Highland (2009) report that 17 % of the fatalities from natural hazards are due to landslides. Few additional studies illustrate the societal relevance of loss of live from landslides at national level. In their time series of landslide fatalities in Italy, Salvati et al. (2010), for instance, record a total of more

Table 1.1 Total annual losses caused by landslides in different countries worldwide

Country	Total annual loss (USD billion)	Loss as percentage of GDP
USA	2.1–4.3	0.01–0.03
Japan	>3.0	>0.06
Italy	3.9	0.19
India	2.0	0.11
China	>1.0	0.01
Germany	0.3	0.01

The cost estimates include direct and indirect losses and are also shown as percentage of national GDP (Gross Domestic Product). Note that the presented losses are rough estimates that only give a first impression of the overall economic significance of landslide impact
Data modified after Li (1989), Schuster (1996), Schuster and Highland (2001), and Klose and Damm (2014). Losses presented in the table are given in 2014 values. National GDPs according to International Monetary Fund (https://www.imf.org/external/data.htm)

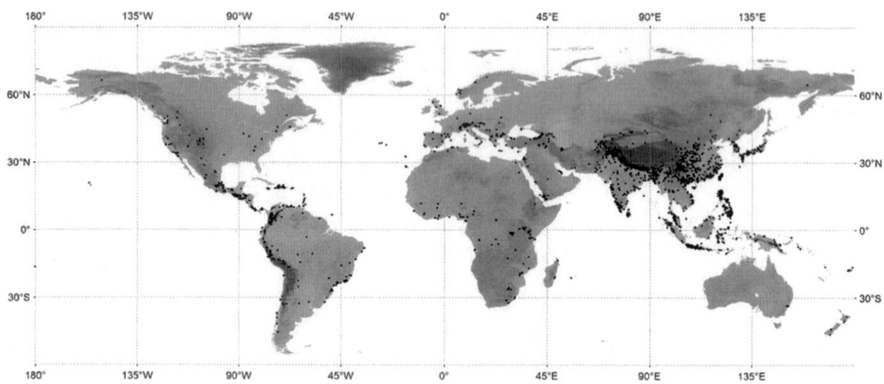

Fig. 1.3 Global distribution of fatal landslides in the period 2004–2010. The map illustrates the location of 2,620 landslides that caused the death of 32,322 people. Regional clusters of fatal landslides are found to be in E- and SE-Asia, the Himalayas, and Central America. Note that this map ignores loss of life from earthquake-induced landslides (*Source* Petley 2012)

than 15,000 deaths between the mid-8th century and 2008. However, the most fatal series of landslides ever recorded was related to the 1920 Haiyuan earthquake that caused the death of approximately 100,000 people (Close and McCormick 1922; see also Runqiu 2009).

Landslides result in economic losses that are both direct and indirect and that are either of public or private nature. Direct losses are the costs of repair, replacement, and maintenance of property damage within the boundaries of landslides. All other losses (property devaluation, loss of revenue, etc.) are indirect, including those caused off-site by secondary hazards such as flooding (cf. Fleming and Taylor 1980; Schuster and Fleming 1986; Schuster 1996). As avoidance costs often relate to post-disaster mitigation (Klose et al. 2014a), the costs of preventing landslide damage are sometimes also classified to direct losses (e.g., Klose et al. 2014b). While direct losses are capped according to landslide magnitude or local setting, those of indirect nature show broader economic impact, especially through complex multiplier effects. For example, one of the costliest landslides of all times is the 2013 Bingham Copper Mine landslide (Utah, USA); a landslide that caused estimated production losses of around US$700 million, with the potential to increase global copper prices (The Landslide Blog 2013; see also Pankow et al. 2014; Fig. 1.4c). An overview on the different types of indirect losses is given by, amongst others, Sidle and Ochiai (2006), Kjekstad and Highland (2009), and Alimohammadlou et al. (2013). Furthermore, the differentiation in public and private losses is crucial for risk assessment, highlighting that the financial burdens of landslides are often not allocated to the general public but paid by affected private property owners alone (cf. Fleming and Taylor 1980; Schuster and Fleming 1986; Schuster 1996). This especially applies to the many cases with absence of landslide insurance (e.g., Olshansky 1996).

From a global perspective, Italy (US$3.9 billion) and Japan (>US$3.0 billion) are among those countries that experience the worst economic impact of landslides worldwide (Table 1.1; see also Trezzini et al. 2013). However, it is not just Italy and Japan where total annual landslide costs amount to billions of dollars but the USA, China, and India as well. The direct costs of landslides in Germany have first been estimated by Krauter (1992) who presented an estimate of US$250 million. Given the latest research results (Sect. 5.3), this cost estimation seems too optimistic, wherefore revised upwards to US$300 million in recent studies (Table 1.1; Klose and Damm 2014). Most of the available cost figures are rough estimates of only preliminary character; however, they are still able to reflect the overall significance of landslide losses in realistic ways. When considering the fact that U.S. cost estimates are transferable to similar industrialized nations (Schuster 1996), the annual worldwide costs of landslide impact would roughly be around US$20 billion, which corresponds to 17 % of the 1980–2013 annual average global natural disaster losses (US$121 billion; according to Munich Re 2014).

A number of previous cross-sector loss studies show consensus that the transportation sector is most affected by the billions of dollars in annual landslide losses worldwide (e.g., Wang et al. 2002; Highland 2012; Klose et al. 2012; Vranken et al. 2013; Sect. 4.3.1). For the U.S. state highway system, Walkinshaw (1992), for instance, estimated annual repair and maintenance costs of US$190 million. Alternatively, Klose and Damm (2014) provide a first cost estimate of annual direct losses for highways (i.e., Bundesstraßen; comparable with U.S. routes) in Germany. According to their cost extrapolation, national losses for this category of road range between US$70 and 80 million each year. Besides large amounts of damage and prevention costs, indirect landslide losses, especially due to traffic disruption, are also critical for transportation infrastructures (e.g., MacLeod et al. 2005; Ohara et al. 2008). With regard to railways, the 2013 Hatfield Colliery landslide (South Yorkshire, UK), for example, illustrates that disruption costs could quickly run into tens of millions of dollars (Symes and Madill 2013; see also BGS 2014). Further case studies of indirect losses together with

Fig. 1.4 (continued)

Fig. 1.4 (continued)

(d)

Fig. 1.4 (continued)

Fig. 1.4 (continued)

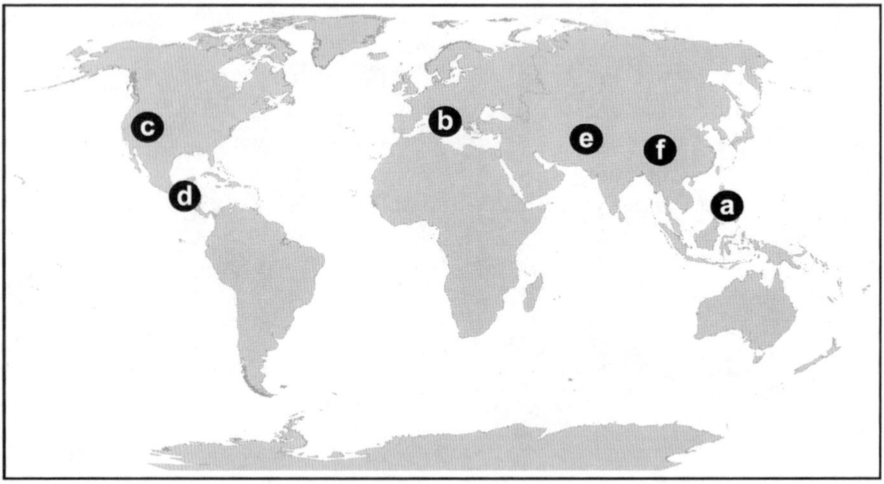

Fig. 1.4 Recent landslide disasters worldwide. **a** Rockslide–debris avalanche on Leyte Island (Central Philippines) in the year 2006. The landslide was caused by tectonic weakening within the Philippine fault zone (*Source* Evans et al. 2007). **b** The 2010 Maierato landslide near the city of Vibo Valentia, Southern Italy, which occurred after a seven-month wet period (*Photo* V. Comerci, ISPRA; cf. Gattinoni et al. 2012). **c** 2013 Bingham Copper Mine landslide (Utah, USA) that is among the most expensive landslides worldwide. The costs due to disruption of mining activity are estimated at about US$700 million (*Photo* Rio Tinto Kennecot; cf. The Landslide Blog 2013). **d** A year-2001 earthquake-induced landslide in El Salvador (Central America) affecting a local community close to the capital city San Salvador (*Photo* E. Harp, USGS). **e** Heavy rainfall caused the 2014 Afghan landslide with possibly more than 2,000 fatalities (*Photo* F. Waezi, UNAMA; cf. Witze 2014). **f** A series of catastrophic landslides triggered by the 2008 Wenchuan earthquake destroyed the town of Qushan in Sichuan Province, SW China (*Photo* D. Wald, USGS)

additional data sets on national cost estimates are provided by Fleming and Taylor (1980), Brabb and Harrod (1989), Schuster (1996), and Schuster and Highland (2001).

Note that all landslide losses presented in this study are given in U.S. dollar and refer to 2013–2014 values. Loss data published in previous studies are adjusted for inflation by using the U.S. consumer price index (CPI-U 08-2013 or CPI-U 05-2014; http://www.bls.gov/cpi/). In the following sections, data sets on landslide losses exclusively relate to direct costs, and the words costs and losses are used as synonyms.

1.3 Research Gap

The nature of landslides is complex in many respects, with landslide hazard and impact being dependent on a variety of factors (cf. Sect. 1.1). This obviously requires an integrated assessment for fundamental understanding of landslide risk. Integrated risk assessment, according to the approach presented in this work, implies combining prediction of future landslide occurrence with analysis of landslide impact in the past. A critical step for assessing landslide risk in integrated perspective is to analyze what types of landslide damage affected people and property in which way and how people contributed and responded to them. As is the case with assessing landslide hazard (cf. Varnes and IAEG 1984; Sect. 4.2.1), the past and the present are also seen as keys to assess landslide impact. In integrated risk assessment, the focus is on systematic identification and monetization of landslide damage, and analytical tools that allow deriving economic costs from physical landslide processes are at the heart of this approach. The broad spectrum of landslide types and process mechanisms as well as nonlinearity between landslide magnitude, damage intensity, and direct costs are some main factors explaining recent challenges in risk assessment. Most previous concepts and models for analyzing landslide risk (cf. Sect. 4.3.1) ignore both the specific characteristics of landslides and the complex nature of landslide impact. It is common practice today to derive landslide risk without considering landslide process-based cause-effect relationships.

The two prevailing approaches for assessing the impact of landslides or related geohazards in economic terms are cost survey (ex-post) and risk analysis (ex-ante). While cost survey has a focus on identifying past landslide losses, risk analysis is trying to predict those occurring in future (e.g., Hallegatte and Przyluski 2010; Meyer et al. 2013; Kreibich et al. 2014). Both approaches are able to complement each other, but yet a combination of them has not been realized so far. Alternatively, new concepts or tools that expand these two approaches are still missing or their potential has only been unlocked to some extent by now. As valuable sources of landslide information, landslide databases (cf. Chap. 2), for instance, play a vital role in both approaches. Thus, cost survey is often a major step in database development, whereas risk analysis takes advantage of parts of their content, especially spatiotemporal landslide data. Landslide databases are therefore also a key component for and a starting point of integrated risk assessment (Fig. 1.5). This approach is based on a systematic framework that combines cost survey and GIS-based tools for hazard mapping or cost modeling with methods to assess interactions between land use and landslides in historical perspective. However, knowing where, when, and why landslides will occur and how much they are likely to cost is only one part of integrated risk assessment. Fundamental understanding of landslide risk also requires knowledge about the economic and fiscal relevance of landslide losses, wherefore analysis of their impact on public budgets is a further component of this approach. In integrated risk assessment, a combination of methods plays an important role, with the objective of collecting and integrating complex data sets on landslide risk.

The research gap in integrated risk assessment becomes apparent when considering the research background of landslide loss studies (Fig. 1.6). From a global perspective, there has been two main eras of cost estimation so far: an empirical era focused on past landslide losses (cost survey), and a modeling era dealing with landslide losses in future (risk analysis). Much of the research during the

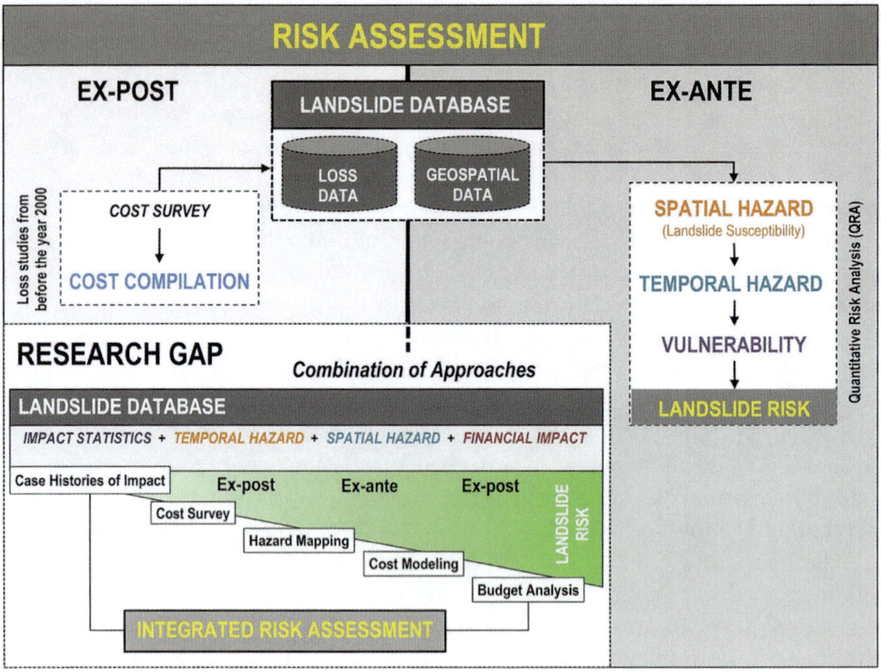

Fig. 1.5 Research gap existing in integrated assessment of landslide risk. The main idea of integrated risk assessment is to combine cost survey (ex-post) with risk analysis (ex-ante) and to make use of landslide databases in impact statistics or in studying the role of human activity in landslide hazard

empirical era was from the U.S. (e.g., Schuster 1996; Sect. 4.3.1) which show the longest history of loss studies. Besides few global reports (e.g., Brabb and Harrod 1989; Schuster and Highland 2001), past landslide losses in Europe, for instance, have only recently received increasing scientific attention, wherefore few additional, non-American cost surveys are available up until now (e.g., Klose et al. 2012; Trezzini et al. 2013; Vranken et al. 2013). Alternatively, the modeling era, including GIS-based risk studies, is dominating since the mid-1990s. The main approaches and methods of this second era are summarized by, amongst others, Dai et al. (2002), Lee and Jones (2004), Crozier and Glade (2005), Chacón et al. (2006), Van Westen et al. (2006), and Corominas et al. (2014). Despite exhaustive research in both eras, cost estimation is still faced with major challenges, which is why some few studies today discuss new integrated approaches (e.g., Crovelli and Coe 2009; Klose et al. 2014b). These recent studies combine cost survey with elements of risk analysis, either to predict future landslide losses from those in the past or to overcome limitation of cost survey to small spatial scales. The methodology of integrated risk assessment proposed in the present work is at the heart of this new development in the evolution of risk assessment approaches. More specifically, this work tries to bridge several research gaps, not only this between cost

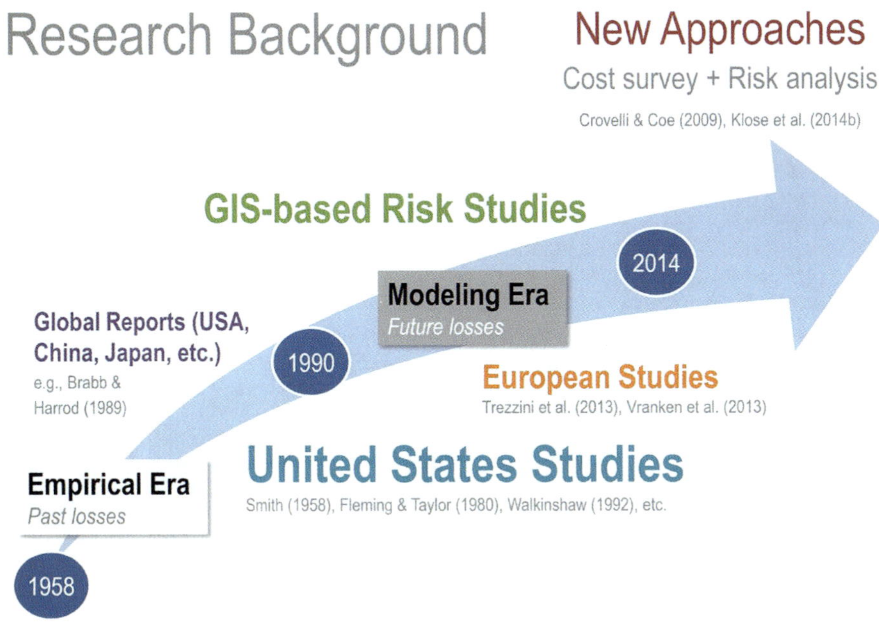

Fig. 1.6 Overview of the research background of landslide loss studies. Since Smith (1958) published first loss data for the U.S., there has been two main eras of cost estimation: an empirical era focused on cost survey (past losses), and a modeling era dealing with risk analysis (future losses). Some few studies today try to close the research gap in integrated risk assessment by presenting approaches that combine cost survey with risk analysis

survey and risk analysis, but also this between landslide process investigation and economic loss study. A key role in addressing these research deficits is played by landslide databases whose potential in impact statistics, hazard mapping, and cost modeling makes them useful tools for integrated assessment of landslide risk (cf. Chap. 2).

This section has outlined the state of research and associated research gaps from an overall perspective. For further information on the approaches and deficits in the different fields of risk assessment, it is referred to the introductory remarks on landslide databases (Sects. 2.1 and 2.2.1). Furthermore, Chap. 4 provides an overview of available methods for assessing temporal or spatial landslide hazard (Sect. 4.1.1 or Sect. 4.2.1) and landslide costs (Sect. 4.3.1).

1.4 Objectives and Outline of the Study

The overarching goal of this study is to develop and apply new approaches that allow using an available landslide database as tool to assess landslide risk in the Lower Saxon Uplands, NW Germany, in integrated perspective. Much of this research is based on a landslide database that has been established in the late 1990s and that is now maintained as national landslide database for Germany (Damm and Klose 2014, 2015). On the basis of the subgoal to contribute to further database development at national level, the present study has the following hierarchically connected priority goals with regard to landslide risk assessment in the Lower Saxon Uplands: (A) modeling of spatial hazard, (B) calculation of temporal hazard, and (C) study of hazard impact (see also Fig. 1.7). The latter priority goal is complemented by three additional goals, including (C.1) impact statistics, (C.2) analysis of hazard management, (C.3) landslide cost modeling, and (C.4) fiscal assessment of cost impact. These priority goals can be differentiated in the following subgoals and working steps:

A. Spatial hazard: ***modeling of landslide susceptibility***

- Development and application of a regional landslide susceptibility model for the German Federal State of Lower Saxony by modifying an available statistical modeling approach
- Mapping of landslide sites in the Lower Saxon Uplands (southern Lower Saxony) by geocoding of landslides stored in a database and creation of a spatial landslide inventory
- Investigation, mapping, and statistical analysis of geofactors that control regional landslide susceptibility and evaluation of the role of human activity in causing landslides

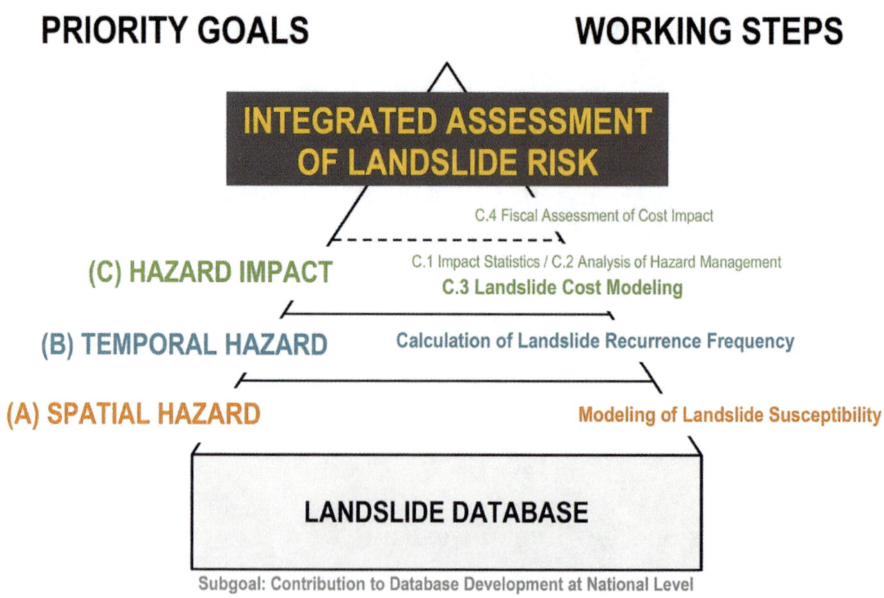

Fig. 1.7 Summary of the priority goals and major working steps of this study

- Application of the landslide susceptibility model for mapping and zonation of landslide-prone areas and identification of regional hazard clusters and key areas of landslide activity
- Mapping of infrastructure area at risk of landslides at local and regional level and assessment of hazard exposure to urban area and road infrastructure in the Lower Saxon Uplands

B. Temporal hazard: ***modeling of landslide-causative soil moisture levels***

- Development of a soil water balance model with regional applicability by combining existing methods for modeling of soil moisture levels connected with past landslide activity
- Model application to simulate monthly soil moisture levels for reference landslide sites in the Upper Weser area, southern Lower Saxon Uplands, since the mid-20th century
- Identification of temporal correlations between simulated soil moisture and landslide activity as well as calculation of critical soil moisture thresholds associated with past landslides
- Calculation of recurrence intervals of landslide-causative soil moisture levels and magnitude–frequency analysis with regard to landslide volume and critical soil moisture levels
- Analysis of soil moisture anomalies to determine the duration of wet periods and soil saturation before landslide occurrence as well as calculation of landslide initiation times

C. Hazard impact: ***impact studies (C.1, C.2, C.4) and landslide cost modeling (C.3)***

C.1. Impact statistics

- Application of the landslide database for analysis of the regional frequency and causes of landslides using the examples of Lower Saxony and the entire German Central Uplands
- Statistical analysis of database contents focused on identification of affected infrastructure, types of damage or disaster response, and methods of landslide repair or hazard mitigation
- Database application to develop case histories of landslide impact for reference sites in the Lower Saxon Uplands to study the role of human activity in the evolution of landslide risk

C.2. Analysis of hazard management

- Database analysis combined with expert interviews and field studies to identify hazard management practices (repair, prevention, maintenance) along highways in Lower Saxony
- Design of disaster management process models for identification and simulation of cost factors involved in coping with landslide hazards at highways over the full disaster cycle

C.3. Landslide cost modeling

- Development of a method for estimation and GIS-based regionalization of direct landslide costs for transportation infrastructures by using a database and the tools from (A) and (C.2)
- Configuration of the method for its application in a loss study for highways in the Lower Saxon Uplands based on the results of the analysis of regional hazard management
- Compilation of landslide costs for a representative case study area over a reliable time period by means of cost survey and cost modeling with a local landslide database subset
- Cost extrapolation for highways in the Lower Saxon Uplands by using a regional infrastructure exposure index and a cost index of the losses for highways in local hazard areas

C.4. Fiscal assessment of cost impact

- Modeling of disaster financing for highways of the Lower Saxony Department of Transportation and for urban infrastructures of the city of Hann. Münden (Upper Weser area)
- Cost survey and expert interviews to estimate landslide losses in Hann. Münden and analysis of financial data at city level and from the Lower Saxony Department of Transportation
- Assessment of the budgetary impact of landslides and analysis of their fiscal relevance for public construction budgets of both authorities in recent years and in the near future

The present study is organized as follows: In Chap. 2, an overview of the evolution of landslide databases worldwide is given, with special focus to the past and current situation in Europe and Germany. The landslide database for Germany, which served as an important information source for this research, is presented next, including a summary of the goals, the structure, and the contents of the database. This chapter also provides examples of regional database application with regard to analysis of landslide frequency, the investigation of landslide causes or triggers, and landslide impact statistics, where each of these applications is focused on the entire German Central Uplands. Subsequently, in Chap. 3, the study area of the present research work, the Lower Saxon Uplands (NW Germany), is introduced. Main emphasis is thereby placed on the description of prevalent landslide types and processes and the soil materials often affected by landslides in this region. In the following, Chap. 4 highlights the approaches and methods that have been developed, optimized, and applied in this study. An introduction preceding each of the three different sections on method development addresses the specific state of research and the existing research limitations and challenges. Based on the arguments for new or improved methodological approaches, the developed analytical tools and methods are presented, including a soil water balance model, a landslide susceptibility model, and a landslide cost assessment model. Chapter 5 then deals not only with the results of local or regional method application, but a discussion of the obtained results is presented herein as well. The different sections of this chapter are published in three accompanying research papers. Finally, Chap. 6 provides a synthesis of the key findings of the conducted studies, while presenting further research results that put these findings into perspective. The proposed framework of integrated landslide risk assessment is used in this chapter to derive an overall qualitative risk estimate for the Lower Saxon Uplands. The strengths and weaknesses of the performed risk estimation are also critically discussed in the synthesis of the present study.

References

Alcántara-Ayala, I. (2014). The spatial-temporal dimensions of landslide disasters. In A. Ismail-Zadeh, J. U. Fucugauchi, A. Kijko, K. Takeuchi, & I. Zaliapin (Eds.), *Extreme natural hazards. Disaster risks and societal implications* (pp. 61–76). Cambridge: Cambridge University Press.

Alexander, D. (1986). Landslide damage to buildings. *Environmental Geology and Water Sciences, 8*(3), 147–151.

Alexander, D. (1989). Urban landslides. *Progress in Physical Geography, 13*(2), 157–189.

Alimohammadlou, Y., Najafi, A., & Yalcin, A. (2013). Landslide process and impacts: A proposed classification method. *Catena, 104*, 219–232.

Antronico, L., Borrelli, L., Coscarelli, R., & Gullà, G. (2014). Time evolution of landslide damages to buildings: The case study of Lungro (Calabria, Southern Italy). *Bulletin of Engineering Geology and the Environment*. doi:10.1007/s10064-014-0591-y.

Bell, R., Glade, T., & Danscheid, M. (2006). Challenges in defining acceptable risk levels. In W. J. Ammann, S. Dannenmann, & L. Vulliet (Eds.), *RISK21 – Coping with risks due to natural hazards in the 21st century* (pp. 77–87). London: Taylor & Francis Group.

BGS. (2014). Hatfield colliery landslide, South Yorkshire, February 2013. British Geological Survey. Retrieved July 15, 2014, from http://www.bgs.ac.uk/landslides/hatfieldFeb2013.html.

Blaschke, P. M., Trustrum, N. A., & Hicks, D. L. (2000). Impacts of mass movement erosion on land productivity: A review. *Progress in Physical Geography, 21*(1), 21–52.

Brabb, E. E. (1991). The world landslide problem. *Episodes, 14*(1), 52–61.

Brabb, E. E., & Harrod, B. L. (Eds.). (1989). *Landslides: Extent and economic significance*. Rotterdam: A.A. Balkema. 385 pp.

Budimir, M. E. A., Atkinson, P. M., & Lewis, H. G. (2014). Earthquake-and-landslide events are associated with more fatalities than earthquakes alone. *Natural Hazards, 72*, 895–914.

Burke, T. J., Sattler, D. N., & Terich, T. (2002). The socioeconomic effects of a landslide in Western Washington. *Environmental Hazards, 4*, 129–136.

Cardinali, M., Ardizzone, F., Galli, M., Guzzetti, F., & Reichenbach, P. (2000). Landslides triggered by rapid snow melting: the December 1996–January 1997 event in Central Italy. In P. Claps, & F. Siccardi (Eds.), *Mediterranean storms, Proceedings of the EGS plinius conference* (pp. 439–448). Cosenza: Bios Publisher.

Chacón, J., Irigaray, C., Fernández, T., & El Hamdouni, R. (2006). Engineering geology maps: Landslides and geographical information systems. *Bulletin of Engineering Geology and the Environment, 65*, 341–411.

Close, U., & McCormick, E. (1922). Where the mountains walked. *National Geographic Magazine, 41*(5), 445–464.

Cornforth, D. H. (2005). *Landslides in practice: Investigations, analysis, and remedial/preventive options in soils*. Hoboken: Wiley. 624 pp.

Corominas, J., van Westen, C., Frattini, P., Cascini, L., Malet, J.-P., Fotopoulou, S., et al. (2014). Recommendations for the quantitative analysis of landslide risk. *Bulletin of Engineering Geology and the Environment, 73*, 209–263.

Crovelli, R. A., & Coe, J. A. (2009). Probabilistic estimation of numbers and costs of future landslides in the San Francisco Bay region. *Georisk, 3*, 206–223.

Crozier, M. J. (1986). *Landslides: causes, consequences & environment*. London: Croom Helm. 252 pp.

Crozier, M. J. (2010). Landslide geomorphology: An argument for recognition, with examples from New Zealand. *Geomorphology, 120*, 3–15.

Crozier, M. J., & Glade, T. (2005). Landslide hazard and risk: Issues, concepts and approach. In T. Glade, M. G. Anderson, & M. J. Crozier (Eds.), *Landslide hazard and risk* (pp. 1–38). Chichester: Wiley.

Crozier, M. J., Preston, N., & Glade, T. (2013). Landslide impacts. In P. T. Bobrowsky (Ed.), *Encyclopedia of Natural Hazards* (pp. 606–610). Berlin: Springer.

Cruden, D. M. (1991). A simple definition of a landslide. *Bulletin of the International Association of Engineering Geology, 43*(1), 27–29.

Cruden, D. M., & Varnes, D. J. (1996). Landslide types and processes. In A. K. Turner, & R. L. Schuster (Eds.), *Landslides: Investigation and mitigation* (pp. 36–75). Transportation Research Board, Special Report 247. Washington D.C: National Academy Press.

Dai, F. C., Lee, C. F., & Ngai, Y. Y. (2002). Landslide risk assessment and management: An overview. *Engineering Geology, 64*, 65–87.

Damm, B., & Klose, M. (2014). Landslide database for the Federal Republic of Germany: A tool for analysis of mass movement processes and impacts. In K. Sassa, P. Canuti, & Y. Yin (Eds.), *Landslide Science for a Safer Geoenvironment, Vol. 2: Methods of Landslide Studies* (pp. 787–792). Berlin: Springer.

Damm, B., & Klose, M. (2015). The landslide database for Germany: Closing the gap at national level. *Geomorphology*. doi:10.1016/j.geomorph.2015.03.021.

D'Elia, B., Picarelli, L., Leroueil, S., & Vaunat, J. (1998). Geotechnical characterisation of slope movements in structurally complex clay soil and stiff jointed clays. *Rivista Italiana di Geotecnica, 3*(98), 5–47.

Dikau, R., Brunsden, D., Schrott, L., & Ibsen, M.-L. (Eds.). (1996). *Landslide recognition: Identification, movement and causes*. Chichester: Wiley. 251 pp.

Dilley, M., Chen, R. S., Deichmann, U., Lerner-Lam, A. L., Arnold, M., Agwe, J., et al. (2005). *Natural disaster hotspots: A global risk analysis*. Washington D.C.: The World Bank. 132 pp.

Evans, S. G., Guthrie, R. H., Roberts, N. J., & Bishop, N. F. (2007). The disastrous 17 February 2006 rockslide-debris avalanche on Leyte Island, Philippines: A catastrophic landslide in tropical mountain terrain. *Natural Hazards Earth System Sciences, 7*, 89–101.

Fell, R. (1994). Landslide risk assessment and acceptable risk. *Canadian Geotechnical Journal, 31*(2), 261–272.

Finlay, P. J., & Fell, R. (1997). Landslides: Risk perception and acceptance. *Canadian Geotechnical Journal, 34*(2), 169–188.

Flageollet, J.-C. (1999). Landslide hazard – A conceptual approach in risk viewpoint. In R. Casale, & C. Margottini (Eds.), *Floods and landslides: Integrated risk assessment* (pp. 3–18). Berlin: Springer.

Fleming, R. W., & Taylor, F. A. (1980). *Estimating the costs of landslide damage in the United States* (21 pp). Geological Survey Circular 832. Arlington: U.S. Geological Survey.

Gattinoni, P., Scesi, L., Arieni, L., & Canavesi, M. (2012). The February 2010 large landslide at Maierato, Vibo Valentia, Southern Italy. *Landslides, 9*, 255–261.

Geertsema, M., Schwab, J. W., Blais-Stevens, A., & Sakals, M. E. (2009). Landslides impacting linear infrastructure in west central British Columbia. *Natural Hazards, 48*, 59–72.

Glade, T., & Crozier, M. J. (2005a). The nature of landslide hazard impact. In T. Glade, M. G. Anderson, & M. J. Crozier (Eds.), *Landslide hazard and risk* (pp. 43–74). Chichester: Wiley.

Glade, T., & Crozier, M. J. (2005b). A review of scale dependency in landslide hazard and risk analysis. In T. Glade, M. G. Anderson, & M. J. Crozier (Eds.), *Landslide hazard and risk* (pp. 75–138). Chichester: Wiley.

González Díez, A., Salas, L., Díaz de Terán, J. R., & Cendrero, A. (1996). Late quaternary climate changes and mass movement frequency and magnitude in the Cantabrian region, Spain. *Geomorphology, 15*, 291–309.

Guthrie, R. H., & Evans, S. G. (2007). Work, persistence, and formative events: The geomorphic impact of landslides. *Geomorphology, 88*, 266–275.

Guzzetti, F. (2005). *Landslide hazard and risk assessment* (389 pp). PhD thesis, Rheinische Friedrich-Wilhelms-Universität Bonn, Mathematisch-Naturwissenschaftliche Fakultät. Retrieved July 4, 2014, from http://hss.ulb.uni-bonn.de/2006/0817/.

Guzzetti, F., Peruccacci, S., Rossi, M., & Stark, C. P. (2008). The rainfall intensity-duration control of shallow landslides and debris flows: An update. *Landslides, 5*, 3–17.

Guzzetti, F., Mondini, A. C., Cardinali, M., Fiorucci, F., Santangelo, M., & Chang, K.-T. (2012). Landslide inventory maps: New tools for an old problem. *Earth-Science Reviews, 112*, 42–66.

Hallegatte, S., & Przyluski, V. (2010). *The economics of natural disasters: Concepts and methods* (29 pp). Policy Research Working Paper 5507. Washington D.C: The World Bank.

Hampton, M. A., Lee, H. J., & Locat, J. (1996). Submarine landslides. *Reviews of Geophysics, 34*, 33–59.

Harp, E. L., Reid, M. E., McKenna, J. P., & Michael, J. A. (2009). Mapping of hazard from rainfall-triggered landslides in developing countries: Examples from Honduras and Micronesia. *Engineering Geology, 104*, 295–311.

Hearn, G. J., Hunt, T., & d'Agostino, S. (2011). Soil slope stabilization. In G. J. Hearn (Ed.), *Slope engineering for mountain roads* (pp. 165–188). Engineering Geology Special Publication 24. London: Geological Society.

Highland, L. M. (2012). *Landslides in Colorado, USA – Impacts and loss estimation for 2010* (49 pp). Open-File Report 2012–1204. Reston: U.S. Geological Survey.

Hilley, G. E., Bürgmann, R., Ferretti, A., Novali, F., & Rocca, F. (2004). Dynamics of slow-moving landslides from permanent scatterer analysis. *Science, 304*, 1952–1955.

Hong, Y., Adler, R., & Huffman, G. (2007). Use of satellite remote sensing data in the mapping of global landslide susceptibility. *Natural Hazards, 43*, 245–256.

Hungr, O., Leroueil, S., & Picarelli, L. (2014). The Varnes classification of landslide types, an update. *Landslides, 11*, 167–194.

Iadanza, C., Trigila, A., Vittori, E., & Serva, L. (2009). Landslides in coastal areas of Italy. In C. Violante (Ed.), *Geohazard in rocky coastal areas* (pp. 121–141). Special Publication 322. London: The Geological Society of London.

Keefer, D. K. (2002). Investigating landslides caused by earthquakes – A historical review. *Surveys in Geophysics, 23*, 473–510.

Kirschbaum, D. B., Adler, R., Hong, Y., Hill, S., & Lerner-Lam, A. (2010). A global landslide catalog for hazard applications: Method, results, and limitations. *Natural Hazards, 52*, 561–575.

Kirschbaum, D. B., Adler, R., Adler, D., Peters-Lidard, C., & Huffman, G. (2012). Global distribution of extreme precipitation and high-impact landslides in 2010 relative to previous years. *Journal of Hydrometeorology, 13*(5), 1536–1551.

Kjekstad, O., & Highland, L. (2009). Economic and social impacts of landslides. In K. Sassa, & P. Canuti (Eds.), *Landslides – Disaster Risk Reduction* (pp. 573–587). Berlin: Springer.

Klose, M., & Damm, B. (2014). Was kosten Schäden durch Georisiken? Eine Einführung in die Ökonomie von Massenbewegungen. *VECTOR – Das Vechtaer Forschungsmagazin, 6*, 19–21.

Klose, M., Damm, B., Terhorst, B., Schulz, N., & Gerold, G. (2012). Wirtschaftliche Schäden durch gravitative Massenbewegungen. Entwicklung eines empirischen Berechnungsmodells mit regionaler Anwendung. *Interpraevent, 12*, 979–990.

Klose, M., Highland, L., Damm, B., & Terhorst, B. (2014a). Estimation of direct landslide costs in industrialized countries: Challenges, concepts, and case study. In K. Sassa, P. Canuti, & Y. Yin (Eds.), *Landslide Science for a Safer Geoenvironment, Vol. 2: Methods of Landslide Studies* (pp. 661–667). Berlin: Springer.

Klose, M., Damm, B., & Terhorst, B. (2014b). Landslide cost modeling for transportation infrastructures: A methodological approach. *Landslides*. doi:10.1007/s10346-014-0481-1.

Korup, O. (2012). Landslides in the Earth system. In J. J. Clague, & D. Stead (Eds.), *Landslides: Types, mechanisms and modeling* (pp. 10–23). Cambridge: Cambridge University Press.

Krauter, E. (1992). Hangrutschungen – Ein Umweltproblem. In: H. J. Matthias, & A. Grün (Hrsg.), *Ingenieurvermessung 92. Beiträge zum XI. Internationalen Kurs für Ingenieurvermessung 2* (S. V4/1–V4/12). Bonn: Ferd. Dümmlers.

Kreibich, H., van den Bergh, J. C. J. M., Bouwer, L. M., Bubeck, P., Ciavola, P., Green, C., et al. (2014). Costing natural hazards. *Nature Climate Change, 4*, 303–306.

Lee, E. M., & Clark, A. R. (2002). *Investigation and management of soft rock cliffs*. London: Thomas Telford. 400 pp.

Lee, E. M., & Jones, D. K. C. (2004). *Landslide risk assessment*. London: Thomas Telford. 464 pp.

Li, T. (1989). Landslides: Extent and economic significance in China. In E. E. Brabb, & B. L. Harrod (Eds.), *Landslides: Extent and economic significance* (pp. 271–287). Rotterdam: A.A. Balkema.

Lu, N., & Godt, J. W. (2013). *Hillslope hydrology and stability*. Cambridge: Cambridge University Press. 458 pp.

MacLeod, A., Hofmeister, R. J., Wang, Y., & Burns, S. (2005). *Landslide indirect losses: Methods and case studies from Oregon* (14 pp). Open-File Report O-05-X. Portland: Oregon Department of Geology and Mineral Industries.

Malamud, B. D., Turcotte, D. L., Guzzetti, F., & Reichenbach, P. (2004). Landslide inventories and their statistical properties. *Earth Surface Processes and Landforms, 29*, 687–711.

Mansour, M. F., Morgenstern, N. R., & Martin, C. D. (2011). Expected damage from displacement of slow-moving slides. *Landslides, 8*, 117–131.

Marano, K. D., Wald, D. J., & Allen, T. I. (2010). Global earthquake casualties due to secondary effects: A quantitative analysis for improving rapid loss analyses. *Natural Hazards, 52*, 319–328.

Masson, D. G., Harbitz, C. B., Wynn, R. B., Pedersen, G., & Løvholt, F. (2006). Submarine landslides: Processes, triggers and hazard prediction. *Philosophical Transactions of the Royal Society A, 364*, 2009–2039.

Meyer, V., Becker, N., Markantonis, V., Schwarze, R., van den Bergh, J. C. J. M., Bouwer, L. M., et al. (2013). Review article: Assessing the costs of natural hazards – State of the art and knowledge gaps. *Natural Hazards and Earth System Sciences, 13*, 1351–1373.

Munich Re. (1999). Naturkatastrophen in Deutschland: Schadenerfahrungen und Schadenpotentiale. München: Münchener Rückversicherungs-Gesellschaft. 99 S.

Munich Re. (2014). NatCatSERVICE. Retrieved July 15, 2014, from http://www.munichre.com/en/reinsurance/business/non-life/natcatservice/index.html.

Nadim, F., Kjekstad, O., Peduzzi, P., Herold, C., & Jaedicke, C. (2006). Global landslide and avalanche hotspots. *Landslides, 3*, 159–173.

Nadim, F., Høydal, Ø., Haugland, H., & McLean, A. (2011). *Analysis of landslides triggered by anthropogenic factors in Europe* (82 pp). SafeLand FP7, Deliverable D1.6. Retrieved August 13, 2014, fromm http://www.safeland-fp7.eu/results/Documents/D1.6.pdf.

Nemčok, A., Pašek, J., & Rybář, J. (1972). Classification of landslides and other mass movements. *Rock Mechanics, 4*(2), 71–78.

Ohara, J., Fujisawa, K., Ikeda, M., Ueno, Y., Kokuryou, N., & Kasahara, R. (2008). Case study of estimation of financial loss by landslide disaster. In *Proceedings of the First World Landslide Forum, Poster Session Volume* (pp. 73–75). November 18–21, 2008. Tokyo, Japan.

Olshansky, R. B. (1996). Financing landslide hazard mitigation in the United States. *Journal of Environmental Planning and Management, 39*(3), 371–385.

Pankow, K. L., Moore, J. R., Hale, J. M., Koper, K. D., Kubacki, T., Whidden, K. M., & McCarter, M. K. (2014). Massive landslide at Utah copper mine generates wealth of geophysical data. *GSA Today, 24*(1), 4–9.

Petley, D. N. (2012). Global patterns of loss of life from landslides. *Geology, 40*, 927–930.

Petley, D. N., Hearn, G. J., Hart, A., Rosser, N. J., Dunning, S. A., Oven, K., & Mitchell, W. A. (2007). Trends in landslide occurrence in Nepal. *Natural Hazards, 43*, 23–44.

Pike, R. J., Howell, D. G., & Graymer, R. W. (2003). Landslides and cities: An unwanted partnership. In G. Heiken, R. Fakundiny, & J. Sutter (Eds.), *Earth science in the city: A reader* (pp. 187–254). Washington D.C.: American Geophysical Union.

Pitilakis, K., Fotopoulou, S., Argyroudis, S., Pitilakis, D., Senetakis, K., Treulopoulos, K., et al. (2011). *Physical vulnerability of elements at risk to landslides: Methodology for evaluation, fragility curves and damage states for buildings and lifelines* (195 pp). SafeLand FP7, Deliverable D2.5. Retrieved August 8, 2014, from http://www.safeland-fp7.eu/results/Documents/D2.5_revised.pdf.

Rossi, M., Witt, A., Guzzetti, F., Malamud, B. D., & Peruccacci, S. (2010). Analysis of historical landslide time series in the Emilia-Romagna region, northern Italy. *Earth Surface Processes and Landforms, 35*(10), 1123–1137.

Runqiu, H. (2009). Some catastrophic landslides since the twentieth century in the southwest of China. *Landslides, 6*, 69–81.

Salvati, P., Bianchi, C., Rossi, M., & Guzzetti, F. (2010). Societal landslide and flood risk in Italy. *Natural Hazards and Earth System Sciences, 10*, 465–483.

Schuster, R. L. (1996). Socioeconomic significance of landslides. In A. K. Turner, & R. L. Schuster (Eds.), *Landslides: Investigation and mitigation* (pp. 12–35). Transportation Research Board, Special Report 247. Washington D.C: National Academy Press.

Schuster, R. L., & Fleming, R. W. (1986). Economic losses and fatalities due to landslides. *Bulletin of the Association of Engineering Geologists, 23*(1), 11–28.

Schuster, R. L., & Highland, L. M. (2001). *Socioeconomic and environmental impacts of landslides in the western hemisphere.* Open-File Report 01-0276. U.S. Geological Survey. Retrieved July 15, 2014, from http://pubs.usgs.gov/of/2001/ofr-01-0276/.

Schuster, R. L., & Highland, L. M. (2007). The Third Hans Cloos lecture. Urban landslides: Socioeconomic impact and overview of mitigative strategies. *Bulletin of Engineering Geology and the Environment, 66*, 1–27.

Sidle, R. C., & Ochiai, H. (2006). *Landslides: Processes, prediction, and land use*. Water Resources Monograph 18. Washington D.C.: American Geophysical Union. 312 pp.

Sidle, R. C., Pearce, A. J., & O'Loughlin, C. L. (1985). *Hillslope stability and land use.* Water Resources Monograph 11. Washington D.C.: American Geophysical Union. 140 pp.

Smith, R. (1958). Economic and legal aspects. In E. B. Eckel (Ed.), *Landslides and engineering practice* (pp. 6–19). Highway Research Board, Special Report 28. Washington D.C.: National Academy of Sciences.

Symes, C., & Madill, P. (2013). Trouble down the line. *New Civil Engineer*. Retrieved July 15, 2014, from http://www.nce.co.uk/features/geotechnical/trouble-down-the-line/8644799.article.

Terhorst, B. (2001). Mass movements of various ages on the Swabian Jurassic escarpment: Geomorphologic processes and their causes. *Zeitschrift für Geomorphologie NF, Suppl.-Bd. 125*, 65–87.

Terzaghi, K. (1950). Mechanism of Landslides. In S. Paige (Ed.), *Application of geology to engineering practice* (pp. 83–123). Berkey Volume. New York: The Geological Society of America.

The Landslide Blog (2013). Is the Bingham Canyon copper mine landslide the most expensive single mass movement in history? In D. Petley (Ed.), *The Landslide Blog*. Retrieved July 15, 2014, from http://blogs.agu.org/landslideblog/2013/04/16/is-the-bingham-canyon-copper-mine-landslide-the-most-expensive-single-mass-movement-in-history/.

Tonini, M., Pedrazzini, A., Penna, I., & Jaboyedoff, M. (2013). Spatial pattern of landslides in the Swiss Rhone Valley. *Natural Hazards*. doi:10.1007/s11069-012-0522-9.

Trezzini, F., Giannella, G., & Guida, T. (2013). Landslide and flood: Economic and social impact in Italy. In C. Margottini, P. Canuti, & K. Sassa (Eds.), *Landslide Science and Practice, Vol. 7: Social and Economic Impact and Policies* (pp. 171–176). Berlin: Springer.

Ugai, K., Yagi, H., & Wakai, A. (Eds.). (2013). *Earthquake-Induced Landslides*. Proceedings of the International Symposium on Earthquake-Induced Landslides, Kiryu, Japan, 2012. Berlin: Springer. 996 pp.

Urciuoli, G., & Picarelli, L. (2008). Interaction between landslides and man-made works. In Z. Chen, J. Zhang, Z. Li, F. Wu, & K. Ho (Eds.), *Landslides and engineered slopes: From the past to the future* (Vol. 2, pp. 1301–1307). London: Taylor & Francis Group.

Van Den Eeckhaut, M., Poesen, J., Vandekerckhove, L., Van Gils, M., & Van Rompaey, A. (2010). Human-environment interactions in residential areas susceptible to landsliding: The Flemish Ardennes case study. *Area, 42*(3), 339–358.

Van Westen, C. J., Van Asch, T. W. J., & Soeters, R. (2006). Landslide hazard and risk zonation – Why is it still so difficult? *Bulletin of Engineering Geology and the Environment, 65*, 167–184.

Varnes, D. J. (1958). Landslide types and processes. In E. B. Eckel (Ed.), *Landslides and engineering practice* (pp. 20–47). Highway Research Board, Special Report 28. Washington D.C.: National Academy of Sciences.

Varnes, D. J. (1978). Slope movement types and processes. In R. L. Schuster, & R. J. Krizek (Eds.), *Landslides: Analysis and Control.* (pp. 11–33). Transportation Research Board, Special Report 176. Washington, D.C.: National Academy of Sciences.

Varnes, D. J., & IAEG. (1984). *Landslide hazard zonation: A review of principles and practice.* Paris: UNESCO. 63 pp.

Vaunat, J., Leroueil, S., & Faure, R. M. (1994). Slope movements: A geotechnical perspective. In R. Oliveira, L. F. Rodrigues, A. G. Coelho, & A. P. Cunha (Eds.), *Proceedings 7th IAEG Congress* (Vol. 3, pp. 1637–1646). Lisboa, Portugal. September 5–9, 1994. Balkema: Rotterdam.

Vranken, L., Van Turnhout, P., Van Den Eeckhaut, M., Vandekerckhove, L., & Poesen, J. (2013). Economic valuation of landslide damage in hilly regions: A case study from Flanders, Belgium. *Science of the Total Environment, 447*, 323–336.

Walkinshaw, J. (1992). Landslide correction costs on US state highway systems. *Transportation Research Record, 1343*, 36–41.

Wang, Y., Summers, R. D., Hofmeister, R. J. (2002). *Landslide loss estimation pilot project in Oregon* (47 pp). Open-File Report O-02-05. Portland: Oregon Department of Geology and Mineral Industries.

Wichter, L. (2007). Stabilisation of old lignite pit dumps in Eastern Germany. *Bulletin of Engineering Geology and the Environment, 66*(1), 45–51.

Wieczorek, G. F. (1996). Landslide triggering mechanisms. In A. K. Turner, & R. L. Schuster (Eds.), *Landslides: Investigation and mitigation* (pp. 76–90). Transportation Research Board, Special Report 247. Washington D.C.: National Academy Press.

Winter, M. G., & Bromhead, E. N. (2012). Landslide risk: Some issues that determine societal acceptance. *Natural Hazards, 62*, 169–187.

Witt, A., Malamud, B. D., Rossi, M., Guzzetti, F., & Peruccacci, S. (2010). Temporal correlation and clustering of landslides. *Earth Surface Processes and Landforms, 35*(10), 1138–1156.

Witze, A. (2014). Afghan landslide was 'an accident waiting to happen'. *Nature News.* doi:10.1038/nature.2014.15158.

WP/WLI – UNESCO Working Party on World Landslide Inventory. (1993). A suggested method for describing the activity of a landslide. *Bulletin of the International Association of Engineering Geology, 47*(1), 53–57.

Chapter 2
Landslide Databases—State of Research and the Case of Germany

2.1 Evolution of Landslide Databases—An Overview

Landslide databases are valuable sources of information for research on landslides, not only in terms of their causes, types, and processes (e.g., Pelletier et al. 1997; Guzzetti et al. 2009; Rossi et al. 2010; Tonini et al. 2013; Hurst et al. 2013), but also the impacts and risks associated with them (e.g., Guzzetti et al. 2003; Hilker et al. 2009; Van Den Eeckhaut et al. 2010; Klose et al. 2014a). A landslide database, often also referred to as landslide inventory, is a systematic collection of information on past landslides (Hervás 2013). Besides some few event-based inventories, for example, those for earthquakes (e.g., Gorum et al. 2011) or rainfall events (e.g., Tsai et al. 2010), most landslide databases today are of historical nature, recording landslides at local to global scale over time (e.g., Malamud et al. 2004; Galli et al. 2008; Guzzetti et al. 2012). The content and completeness of historical databases is varying strongly, mainly as a function of spatial and temporal data coverage (cf. Van Den Eeckhaut and Hervás 2012a). Global inventories give a valuable overview on distribution patterns and impacts of catastrophic landslides (e.g., Petley et al. 2005; Kirschbaum et al. 2010; Petley 2012; USGS 2014), but as the majority of landslides are local events, rarely receiving worldwide attention, they include in general only a fraction of the many landslides occurring each year (cf. Spizzichino et al. 2010). This is the same for today's natural disaster databases (e.g., CRED 2014; Munich Re 2014) that record just some of the major landslide events worldwide.

A more reliable record of past landslides is usually provided by national or regional landslide databases. Over the past two decades, there has been considerable progress in the development of national landslide databases across the globe (e.g., Glade and Crozier 1996; Devoli et al. 2007; Osuchowski 2008; Liu et al. 2013), especially in many European countries (cf. Dikau et al. 1996; Van Den Eeckhaut and Hervás 2012a). Various studies have recently reported on the structure, content, and application of the 22 national landslide databases existing in Europe today, including, amongst others, Jelínek et al. (2001), Creighton (2006), Komac et al. (2007), Trigila and Iadanza (2008), Jaedicke et al. (2009),

M. Klose, *Landslide Databases as Tools for Integrated Assessment of Landslide Risk*, Springer Theses, DOI 10.1007/978-3-319-20403-1_2

Schweigl and Hervás (2009), Foster et al. (2012), Damm and Klose (2014, 2015), and Mrozek et al. (2014). In Europe as a whole, and in Italy or Germany in particular, regional landslide databases cover either administrative units or specific mountain areas. Most of them are operated for the purpose of analysis of landslide impact, hazard, and risk, including those databases, for example, that have been applied in studies for the Umbria region, central Italy (e.g., Galli and Guzzetti 2007), the Arno River basin, central Italy (e.g., Catani et al. 2005), and the Lower Saxon Uplands, NW Germany (e.g., Klose et al. 2014b, c). An EU-wide overview on regional landslide databases maintained by provincial governments or research institutes is given by Van Den Eeckhaut and Hervás (2012a).

The recent survey by Van Den Eeckhaut and Hervás (2012a) also deals with the content and characteristics of European landslide databases. According to the survey results, most regional and national databases in Europe store besides data sets on core attributes (e.g., location, occurrence date, movement type) a broad spectrum of additional data, ranging from landslide processes (size, velocity, etc.) and triggering or controlling factors (e.g., geology, land use, rainfall) to impact and mitigation of landslides (damage, fatalities, costs, etc.). The level of detail and data completeness, however, differs strongly between available databases, especially regarding additional data. Thus, more than half the databases store such additional data with less than 25 % completeness, whereas spatiotemporal information is frequently included. Much of the available databases cover a time span of the previous 100–1,000 years and contain between several hundred to several ten thousand data sets. Almost every database exists in digital format, with software or database management systems such as ArcGIS, MapInfo, MS Access, and Oracle Database being most frequently applied. A large number of databases, especially those with national focus, provide landslide data online, sometimes by means of a web GIS application (see also Spizzichino et al. 2010; Van Den Eeckhaut and Hervás 2012b). As a closer look on the database websites indicates, data availability and knowledge transfer is frequently limited, which is mainly because of restricted online access, technical problems, and language barriers. Data collection in most cases is primarily based on data mining of press or historical archives, field work, and analysis of a variety of remotely sensed data (e.g., aerial photography, satellite imagery, LiDAR DEMs) (cf. Guzzetti et al. 2012; Van Den Eeckhaut and Hervás 2012a). An increasingly important role in tracking current landslides in Europe or other parts of the world is also played by public participation via online report systems (cf. Baum et al. 2014) or by tools to explore web and social media contents (cf. Battistini et al. 2013).

The Federal Republic of Germany joined only recently the group of EU member states that have available a national landslide database (cf. Damm and Klose 2014, 2015). With the launch of a national database initiative in recent years, a significant step has been made to close the gap at national level that existed in Germany for more than 40 years. Initial efforts in landslide mapping began as early as the mid-20th century (e.g., Ackermann 1959), with the first spatial inventory having been compiled for the Weser-Leine Uplands, NW Germany, by Schunke (1971). However, it was not until the mid-1990s that research projects such as MABIS (Mass Movements in South, West, and Central Germany, 1995–2001) were focused on targeted database development, especially at local and regional level (cf. Dikau and Schmidt 2001). This and more recent projects resulted in landslide databases for different regions in Germany, including Rhine Hesse (e.g., Dikau et al. 1996; Glade et al. 2001), the Bonn metropolitan area (e.g., Grunert and Hardenbicker 1991; Hardenbicker and Grunert 2001), Thuringia (e.g., Baum and Schmidt 2001; Schmidt and Beyer 2001, 2003), the Southern German scarplands (e.g., Bibus and Terhorst 2001; Terhorst and Kreja 2009; Jäger et al. 2012), and the Bavarian Alps (cf. Barnikel and Becht 2004). Development of landslide databases in Germany has traditionally been a research focus at university institutes in the field of Geography. Since about the mid-2000s, however, the efforts in database development declined considerably within this discipline, and a consolidation of collected data sets has not been realized so far. By contrast, some of the former landslide databases are no longer maintained, which seriously threatens their persistency. Despite the previous achievements, a continuation of database development has been increasingly neglected in recent years, which stands in contrast with the leading role Geography is expected to play in today's georisk research (cf. Gans et al. 2014).

Inventory of landslides for large regions is also a major research task of most state geological surveys in Germany today. Landslide databases are now available for four German federal states (Bavaria, Rhineland-Palatinate, Hesse, Saxony), while two further states (Mecklenburg-Western Pomerania, Schleswig-Holstein) are maintaining a landslide database for at least parts of the state (see also http://www.bgr.de/geol_la/geol_la.htm). Most of these databases are accessible online and contain geospatial information for several hundred to a few thousand landslides of modern to pre-Holocene age. Profound insight into the structure and content of these landslide databases provide, amongst others, Obst and Schütze (2010), Bock et al. (2012), and Kött et al. (2012). The research activity at state level is accompanied by some first database initiatives in related disciplines, especially in fields such as transportation planning and coastal management (e.g., Krauter et al. 2012; LKN-SH 2014).

Among the many landslide databases in Germany that have been developed until today, there is still only one database that has a broader geographic and thematic coverage. The database "Landslides in Low Mountain Areas of Germany" established in the late 1990s has been permanently updated and expanded, wherefore serving as an ideal starting point for launching a national database project for Germany at this time. Today, data sets on more than 4,200 landslides with over 13,000 single data files are stored in this database that covers besides the Central Uplands several main distribution areas of landslides in Germany, including the Southern German Scarplands, the Alpine Foreland, and the coasts of the North and Baltic Sea (Fig. 2.1). The timeframe of the database is about the past 150–200 years, with the oldest landslide, however, being recorded

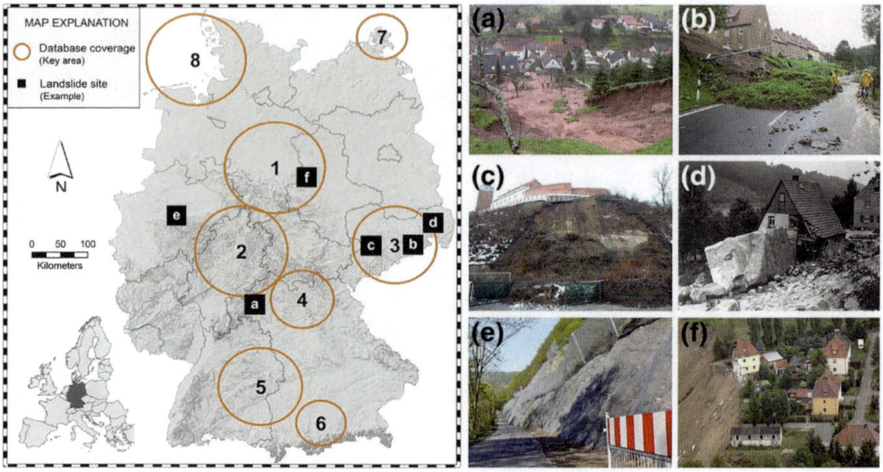

Fig. 2.1 Spatial coverage and key areas (*1–8*) of the landslide database for the Federal Republic of Germany (*Source* modified after Damm 2005). The figure also shows exemplary landslide sites from different parts of Germany: **a** a year-2002 flowslide in highly saturated soil, Neustadt am Main, Bavaria (*Photo* M. Nätscher and R. Stein, THW); **b** rotational slide after intense rainfall at a road cut in Glashütte, Saxony, in the year 2002 (*Photo* H. Weber, Cunnersdorf, Saxony); **c** 2011 Burgberg landslide caused by the collapse of a retaining wall at a cultural heritage site in Eilenburg, Saxony (*Photo* Database B. Damm); **d** historic rockfall (year 1936) near Postelwitz-Schmilka, Saxony (*Source* LfULG); **e** recent landslide mitigation along the Hengsteysee-Trail in Syburg, North Rhine-Westphalia (*Photo* Database B. Damm); **f** 2009 Nachterstedt landslide developed in the overburden of a coal mine in Nachterstedt, Saxony Anhalt (*Photo* Database B. Damm)

as early as 1137. The database takes account of all types of landslides, especially slides and falls, and considers landslides in both urban and rural areas (cf. Damm and Klose 2014, 2015).

The vital role of landslide databases in mapping landslide susceptibility, hazard, and risk has made their development a major research task in Europe and worldwide for the last two decades (e.g., Hervás and Bobrowsky 2009; Spizzichino et al. 2010; Van Den Eeckhaut and Hervás 2012a). Despite being on the top of today's research agenda, the full scientific potential of landslide databases still has only been unlocked to some extent. This especially applies to the use of landslide databases in analysis and statistics of landslide processes, causes, and impacts (cf. Damm and Klose 2014, 2015; Klose et al. 2014a, d). A main reason for this research deficit most likely resides in underestimation of the quality and power of available landslide data, thus the data capabilities of landslide databases for integrated risk assessment. The next section gives an example of database development for such purposes by presenting impact statistics derived from the German landslide database while highlighting its research strategy, structure, and contents.

2.2 Landslide Database for the Federal Republic of Germany

2.2.1 Background and Goals of the Database

The landslide database applied and analyzed in the different studies of this research work was established as early as the late 1990s (cf. Damm and Klose 2014, 2015). Starting from the Upper Weser area, Lower Saxon Uplands (NW Germany), collection and inventory of landslide data has been regionally expanded over time, first to adjacent areas, especially northern Hesse, Thuringia, and eastern Westphalia, and then to selected regions throughout the entire German Central Uplands. The reason to start with the creation of a landslide database in the Upper Weser area by the end of the 1990s was a clustering of landslides in this region at this time. Over the years, landslide data have been gathered for different low mountains areas in Germany, including parts of Saxony, northern Bavaria, and Wuerttemberg. With data mining of web resources complementing field and archive studies since about the mid-2000s, tracking of recent landslide events has become easier and more effective, even over large geographic areas in Germany. From then on, the database has no longer been a compilation of landslide data from selected regions, but rather an inventory of national coverage, at least with regard to most recent landslides. Permanent data collection and update of the database throughout large parts of the country for the last 15 years resulted in a landslide database that is now the most comprehensive for Germany by content and number of recorded landslides (Damm and Klose 2014, 2015).

Major purpose of this landslide database is to store and provide detailed scientific data on landslides in Germany. The database has been developed for studying different aspects of landslides, especially their processes, causes, and impacts, not only at local or regional level, but also over broader geographic areas. While having evolved to a national database in recent years, the database undergoes a far-reaching transformation process today, with the following goals being at the heart of the current research strategy (cf. Damm and Klose 2015):

(i) **Unlocking the full database potential**. The purpose of this goal is to make use of the broad spectrum of methods for database development in order to apply the database in fields in which the potential of landslide databases has long been underestimated. A key to database application in research on landslide processes, causes, and impacts is seen in a targeted strategy of systematic data retrieval. Nowadays, best practices in data retrieval refer not only to mapping approaches, including, for example, analysis of satellite or LiDAR imagery, but also, and more importantly, data mining of the growing pool of landslide data in web, press, and agency archives. To create landslide databases useful for addressing various research questions, systematic data retrieval, however, is only one aspect. The other aspect is information extraction, which is to identify and separate structured information of the collected data material. This step requires approaches capable to search the database for valuable but often hidden information, where in case of archive data, qualitative or expert-based methods, for instance, text analytics, are of growing importance. Landslides are complex phenomena, driven by many different factors, which is why some of their characteristics only become accessible when combining various types of data. A concept of systematic data integration that involves fusion and joint processing of geospatial, geotechnical, and socioeconomic data is therefore a further key to create databases with a broad potential of application. In case of systematic retrieval, extraction, and integration of data from multiple sources, landslide databases show the potential to open a whole new window on the study of landslide processes, causes, and impacts, even at large spatial scales.

(ii) **Data sharing and knowledge transfer**. The goal is to update the national database with as much landslide data as publicly available in Germany in order to create a large data pool that combines the available data in synergistic ways. Integration and centralized storage of the large number of local and regional data sets has not been done so far, despite the large scientific potential a pooling of data would unleash. By contrast, some of these databases are no longer maintained, which seriously threatens their persistency. The recent database migration to PostgreSQL/PostGIS, a high-performance spatial database system, constitutes a milestone to enable consolidation of data sets. Besides having a focus on generating data synergies at national level, the project intends to contribute to European efforts in promoting interoperability of member state databases. High priority is therefore placed on harmonization of data formats and classification systems for ensuring compatibility of the database with those from other European countries. The goal of this initiative is to have available a database for Germany that functions as a tool for data sharing within an evolving EU-wide database network. In addition to support scientific exchange, the role of the database is to serve as a basis for establishing national partnerships with landslide practice. Research activity over the past years already included fruitful cooperation with partners from specialized agencies, especially with ones from transportation and urban planning departments. The objective of expanding such partnerships refers not only to acquisition of first-hand data material, but also, and importantly, to provide professionals with expertise for decision making. A key to knowledge transfer and data sharing is the development of a web GIS application as database frontend and platform for distribution and exchange of information.

2.2.2 Structure, Content, and Information Sources of the Database

A simple file-based system of data storage in the form of a catalog of various directories with a MS Excel database at its heart is defining the basic framework of the landslide database just prior to its current migration (Fig. 2.2). The MS

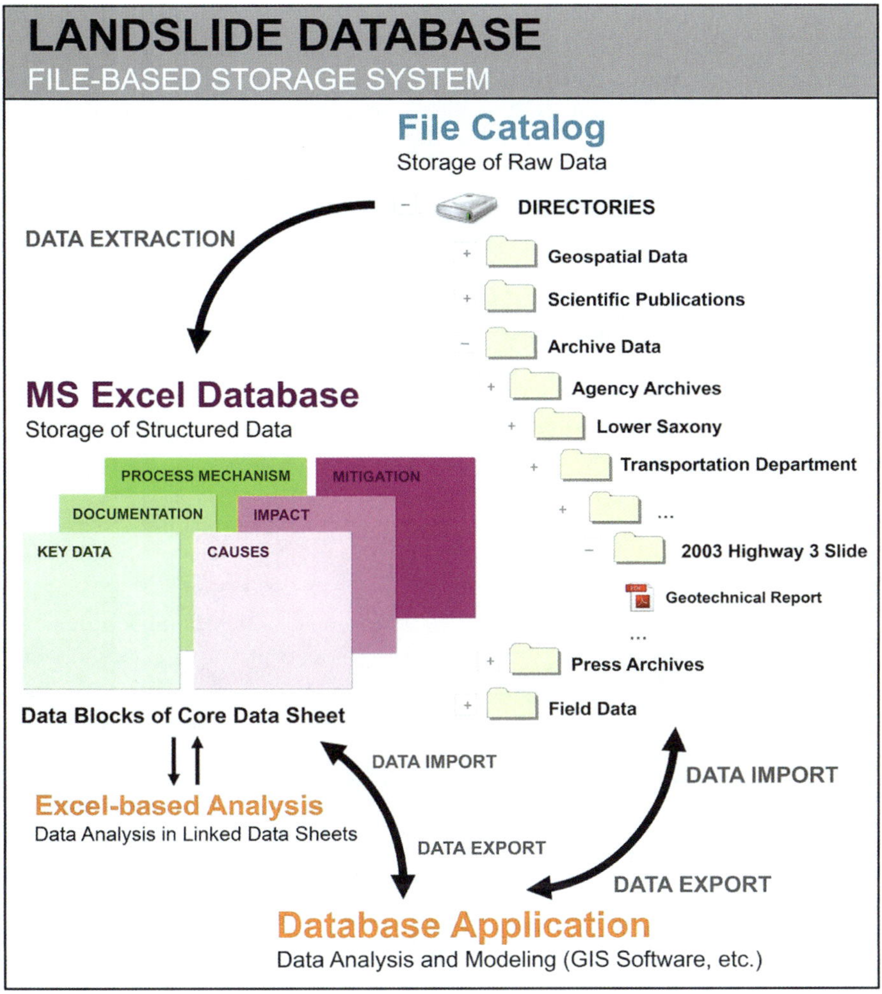

Fig. 2.2 Simplified model of the existing database architecture with the main components of this file-based storage system. Data management is organized on the basis of a file catalog that stores raw data in a system of various directories and data folders as well as a MS Excel database for storage of structured data. For data analysis and modeling, database contents are exported to standard application software. The storage of developed data products is realized either in the MS Excel database or the file catalog (*Source* Damm and Klose 2015)

Excel database only stores structured data (flat file database), which are data that passed through information extraction. All other data sets included in the database, for example, geospatial data (shapefiles, etc.) or climate records (textfiles, etc.), are organized in separate data folders embedded in different directories of the catalog. The data stored in the MS Excel database are arranged in a system of data sheets of which one is the core data sheet that provides a summary of information.

This core data sheet registers chronologically a complete profile for each recorded landslide in tabular format. It is differentiated into seven major data blocks that represent various thematic fields and that include a series of data tables storing numerical data or text information. Besides a data block of KEY DATA and storage space for a LANDSLIDE DOCUMENTATION, there are several data blocks with focus on process- and impact-related aspects of landslides. The segmentation in aggregated data blocks, thematically related data tables, and single data fields is characterizing the structure of the MS Excel database at the moment (cf. Damm and Klose 2014, 2015).

The core attributes of landslides, including identification number, occurrence date, location, administrative region, and data sources, are stored in the KEY DATA for each recorded landslide. These basic identifiers are complemented by a LANDSLIDE DOCUMENTATION providing a textual description of every landslide that is extracted of the entirety of collected data. In addition to a documentation specifying process mechanisms and causes, the data block contains, where applicable, a damage profile that takes account of repair and mitigation measures. This information is completed by the results of a quality assessment for the different types of analyzed data. The data block LANDSLIDE PROCESS MECHANISM contains a collection of multiple data tables, including those labelled as type–depth–size, velocity–magnitude, and activity–stability. Information on predisposing and triggering factors, by contrast, are stored in the data block LANDSLIDE CAUSES. The two main tables of this data block refer either to natural or human factors, and their data categories, for instance, relate to factors such as rock strength, rainfall, and slope modification. A peculiarity of the database is to store also damage information and data on hazard mitigation. As a consequence, the data block LANDSLIDE IMPACT provides besides various data sets on type and severity of damage to infrastructure or mobile objects, detailed information about casualties and fatalities. Alternatively, the data block LANDSLIDE MITIGATION addresses site management, landslide repair, and hazard prevention, thus containing, for example, data sets on methods used for slope stabilization. The available data on economic impact, more specifically damage or prevention costs, are kept separately in the data block LANDSLIDE LOSS.

This database configuration guarantees logic and consistent data storage and provides options for statistical or GIS-based data analysis. Basic statistics (frequency tables, etc.) are usually performed in data sheets connected with the core data sheet through automatized data relations. For advanced analysis of database contents (e.g., cost modeling, see Sect. 4.3.2), however, relevant data sets are extracted to separate Excel files. Statistical analysis of data is mostly performed on the basis of predefined or customized functions of the Excel formula library and specially developed calculation tools. Alternatively, storage and processing of spatial data sets is done by using GIS software, where ArcView GIS and SAGA GIS have frequently been used in previous studies (e.g., Varga et al. 2006; Damm et al. 2010; Klose et al. 2014c). The storage of developed data products is realized either in the MS Excel database or in the file catalog. Most data products are available in form of time series, index or threshold values, data tables, diagrams, and maps (cf. Damm and Klose 2014, 2015).

The landslide database stores data sets derived from a variety of different information sources (Fig. 2.3). More than 40 % of the ~3,000 information sources that have been analyzed to develop the database are characterized by a high level of reliability, including types of sources such as scientific publications (23 %), field data (6 %), and agency archives (14 %). With regard to information from agency archives (transportation departments, etc.), building files, geotechnical reports, and maintenance protocols show the highest information content. These information sources are capable to provide highly valuable data, ranging from landslide material and process mechanism to types of landslide damage and mitigation. Most of these data sets are available in paper or digital format and are usually stored in in-house archive systems. Fire departments or the Federal Agency for Technical Relief (THW), however, increasingly release disaster information in publicly accessible online databases as well.

A total of 12 % of the information sources constitute geospatial data products, for instance, published maps, satellite data, and press photography (Fig. 2.3). In database development, these sources, especially former landslide inventory maps or Google Earth imagery, were used to determine landslide location or basic landslide features. Alternatively, data on local setting and geoenvironmental conditions

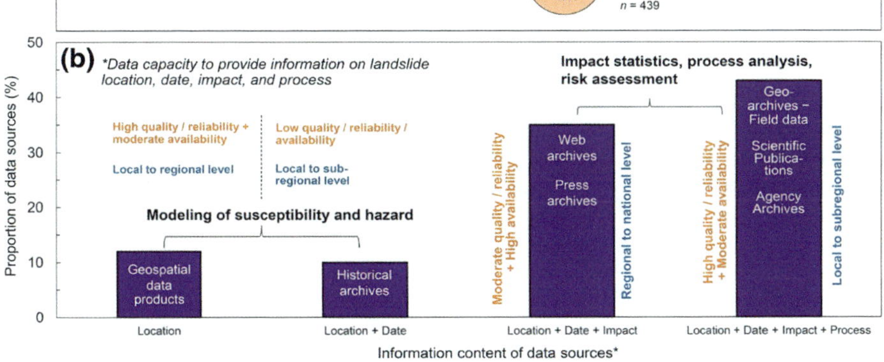

Fig. 2.3 Information sources of the database (**a**) and their capacity to provide data on landslide location, date, impact, and process (**b**). The information content of the different source categories is evaluated on the basis of a qualitative assessment of the average level of detail in their data. It is important to note that within each category data quality and reliability is often strongly varying as a function of source origin (*Source* Damm and Klose 2015)

is extractable of thematic and historical maps (lithology, vegetation, land use, etc.) that are often accessible via online map viewers. This category of geospatial data also includes press and historical photos or multi-year photo collections from mitigation projects, with both of which serving as a good basis for gathering impact-related data.

A further group of information sources are press archives (24 %) to which online access is often guaranteed in recent years (Fig. 2.3). The data quality of press or traffic reports and newspaper articles is usually good enough to provide some basic information on landslide location and date and to some extent landslide impact as well. Main advantage of these types of sources is their capacity to provide such data with a high level of spatiotemporal availability. Data completeness and quality, however, strongly depends on the origin of information, wherefore not every source is equally suited for impact research. Despite some justified criticism, data from press or web archives are important reference points on landslide occurrence, thus being a key to access detail data in agency databases.

As compared with press archives, quality and reliability of data from historical archives, which are the smallest group of information sources (10 %), is significantly lower, enabling only estimation of landslide location and date (Fig. 2.3). The value of sources from pre-modern age, for example, chronicles and annals, is rarely of such good quality to support extraction of clearly datable and locatable landslide data. As is the case with press archives, historical data sets are characterized by a large variation in data quality and completeness, which mainly relates to author experience. However, quality of historical sources increased over time, and data sets showing the level of accuracy required in statistical data analysis usually became available since the mid-19th century (cf. Damm and Klose 2015).

The development of the database is based on a top-down approach of data retrieval that combines broadening of data coverage at national level with local data specification. According to this approach, data collection usually starts with systematic web content mining and analysis of online emergency databases. Gathering of landslide information at national level is assisted by the use of web alerts and tools for web monitoring that enable landslide news tracking over broad areas. The objective of local data specification is to add detail to the data pool in areas that show high landslide density. In these cluster areas, further archive studies are performed, whereby the selection of archives follows spatial and thematic aspects. Most of the relevant information is usually stored in state, county, and city archives or archives of transportation, forest, and urban planning departments.

The starting point of archive studies are often press reports that serve as first pieces of information for archive selection and the search in archives. Despite increasing availability of digital database systems, archive studies are still a time consuming task, with a major obstacle being related to finding the right identifiers used in archive databases. The purpose of field studies, by contrast, is to prove and validate the collected data on a case-by-case basis and to fill data gaps if necessary. In the field, there is only time to collect basic data on landslide size and local setting; nevertheless, the database integrates results of detailed field investigations as well. Finally, analysis of geospatial data products is an important task of this top-down approach, but one that takes place during the entire process of data collection (cf. Damm and Klose 2014, 2015).

2.2.3 *Examples of Regional Database Application*

2.2.3.1 Analysis of Regional Landslide Frequency

The most complete data set of dated landslide information in the database is recorded for the German Central Uplands, including regions such as eastern Westphalia, northern and eastern Hesse, southern Lower Saxony, and western Thuringia (cf. Fig. 2.1, "area 1" and "area 2"). A landslide time series extracted from this database subset includes a total of 1,720 landslide events between 1820 and 2013 (Fig. 2.4a). Landslide activity in the Central Uplands shows a strong increase over this period of time, while the long-run trend, however, is superimposed by strong fluctuation in annual frequency of landslides. Partitioning of the time series reveals that annual mean landslide frequency for some main eras of information availability ($\leq$1869, 1870–1949, 1950–1999, $\geq$2000) is rising significantly over time. More specifically, it increases almost 50-fold over the whole observation period, from a value of 0.9 for the 1820–1869 period ("pre-newspaper era") to a value of 44.6 for the 2000–2013 period ("internet era"). The time series shows thus a strong trend component, but there is no indication that this trend has a cyclic behavior. According to Fig. 2.4a, periods of above- or below-average landslide frequency are alternating irregularly, and much of the annual variability in landslide frequency is proven to be related to random fluctuations. Variation in the annual number of landslides differs throughout the time series, being highest for the 2000–2013 period ($\sigma = 24.7$) as indicated by a comparison of standard deviations (cf. Damm and Klose 2015).

Correlation of landslide activity with the 1901–2010 national rainfall trend (Fig. 2.4b) enables to explain at least some of the variability in annual landslide frequency. There is first evidence for a rainfall pattern in landslide activity, indicating that years with rainfall anomalies (Fig. 2.4c) often showed exceptional landslide activity as well. For example, the wet period 1965–1966 or the year 2002, which had similar record rainfall (976 mm), both correspond to peaks in annual landslide frequency (e.g., 1965, 38 landslides; 2002, 79 landslides). Dry periods such as the year 1991 (644 mm), by contrast, are connected with low landslide frequency (3 landslides), even though absolute deviation is less pronounced on average. The scatterplot illustrated in Fig. 2.4a describes this positive correlation between annual precipitation and number of landslides per year; however, the strength of the correlation as measured by Pearson's r is found to be of only low to moderate intensity ($r = 0.32$).

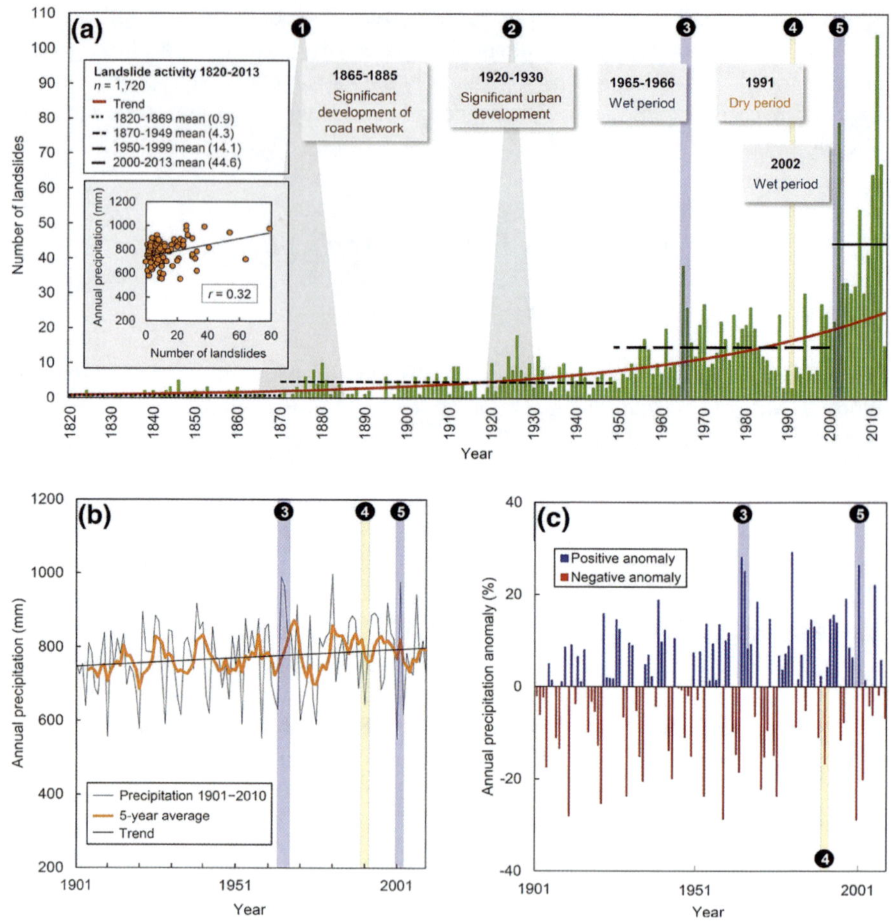

Fig. 2.4 Landslide activity in the German Central Uplands in the period 1820–2013 (**a**) and 1901–2010 annual trend (**b**) or anomaly (**c**) in national average precipitation (rainfall data according to Schönwiese 2013). The strong increase in the number of recorded landslides over time is mainly related to continuously improved data availability. Some of the annual variability in landslide frequency can be correlated with periods of significant infrastructure development (*1–2*) and positive or negative annual rainfall anomalies (*3–5*). The correlation analysis in (**a**) proves that the overall influence of rainfall on landslide activity is less significant ($r = 0.32$) on an annual basis (*Source* Damm and Klose 2015)

2.2.3.2 Analysis of Causes and Triggers of Landslides

The database subset of landslides in the Central Uplands has been used to analyze predisposing (1138 landslides) and triggering factors (680 landslides) (Fig. 2.5). Most landslides recorded between 1960 and 2013 were caused by a combination of causative factors. As complex factor interaction makes it difficult to determine statistically the role of each factor, the presented statistics are based on a

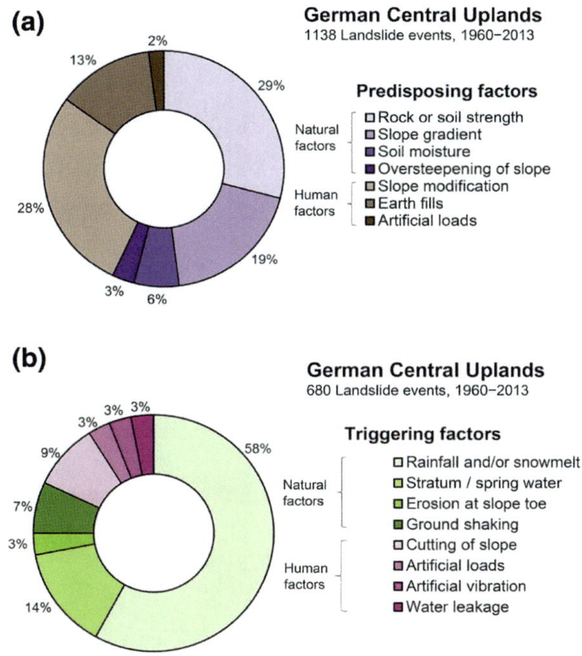

Fig. 2.5 Predisposing (**a**) and triggering factors (**b**) of landslides in the German Central Uplands between 1960 and 2013. Most landslides were caused by a combination of causative factors. The statistics are therefore based on a qualitative evaluation about which factor showed most likely the strongest destabilizing effect. High soil moisture levels as a result of prolonged wet periods as well as construction works constitute the most relevant predisposing factors. A large number of landslides are strongly associated with climatic triggering events such as intense rainfall and/or rapid snowmelt (*Source* Damm and Klose 2015)

qualitative evaluation about which factor showed most likely the strongest destabilizing effect. It is important to note that there are often problems to clearly differentiate between predisposing and triggering factors, wherefore these statistics are fraught with considerable uncertainty. Despite the simplification made in this study, the analysis verifies that landslides as a whole are controlled by a variety of different predisposing or triggering factors, both of which either of natural or anthropogenic nature.

Of the more than 1,100 landslides showing a reliable record, 57 % were related to natural predisposing factors, where in as many as 43 % of the cases, human activity was found to be the main reason for reducing slope stability (Fig. 2.5a). Besides strength properties of rock or soil (29 %), predisposing factors such as slope gradient (19 %), soil moisture (6 %), and oversteepening of slope (3 %) were of high relevance as well. Alternatively, a major role in destabilization of slopes was played by various human activities, including slope modification (28 %), construction of earth fills (13 %), and artificial loads (2 %). A large part of landslides in the Central Uplands were rainfall-induced, wherefore soil saturation by intense precipitation (58 %) is seen as key triggering factor, often in combination with rapid snowmelt (Fig. 2.5b). Further triggering factors of natural origin are partly difficult to categorize, including erosion at slope toe (3 %), spring discharge (14 %), and ground shaking (7 %). Human-triggered landslides were usually the result of construction works, especially activities such as slope cutting (9 %), heavy loads (3 %), and vibration (3 %); but sometimes they were also connected to uncontrolled water leakage (3 %). The triggering of landslides in this part of Germany has thus mostly a climatic reason, with about 60–70 % of past landslides having been directly influenced by rainfall and/or high soil moisture levels (cf. Damm and Klose 2015).

2.2.3.3 Regional Statistics of Landslide Impact

Using the example of the 1960–2013 landslide sample for the Central Uplands (see above), statistical analyses were conducted to study the impact of landslides on people and infrastructure at regional level. The statistics presented in Fig. 2.6a show that landslides primarily affected traffic routes, with roads (37 %) and railways (14 %) having been most often involved in damage events. Much of the landslides along traffic routes were shallow soil slides or rockfalls and resulted in types of damage that range from burial to structural damage. By contrast, damage to buildings (19 %), especially private homes, was frequently related to slow-moving landslides as well. Severity of building damage had often been a function of time, meaning that hardly visible damage intensified to total loss in the long run. Although with significantly lower frequency, landslides also caused damage to lifelines (4 %), waterways (8 %), and forest or agricultural areas (10 %). Further land use types regularly affected by landslide damage were sports fields, mining areas, graveyards, and sites of cultural heritage. A closer look at the various types of damage shows that at traffic routes, for example, not only failure of cut slopes or embankments, but also collapse or tilting of old (masonry) retaining walls caused frequent problems in the past. Alternatively, building damage was besides crack formation in walls and foundations, mainly related to burial of backyards or collapse of building back walls (cf. Damm and Klose 2015).

The most common approach of disaster response (Fig. 2.6b) was repair and further use of affected infrastructures (85 %), whereas permanent abandonment of use (15 %) played a major role only at private homes or other non-commercial buildings. Once affected by landslide damage, more than 80 % of the buildings were vacated over time, which was mainly due to loss of structural integrity and/or high repair costs. Along traffic routes, by contrast, repair or mitigation of landslide damage was dominating disaster response, although few examples of permanent road closure are documented as well.

In case of hazard mitigation (Fig. 2.6c), simple prevention measures at minimal cost show highest frequency, especially with regard to transportation infrastructure. Three types of prevention measures were of special importance in the past: removal of rock and vegetation (34 %), catch barriers (8 %), and rockfall drapery (12 %). Due to the fact that having often been undersized, catch barriers tended to fail under stress, which frequently caused serious traffic accidents. Up until today, the database records ~90 fatalities and ~150 casualties for this region, and significant damage to vehicles (~70) and trains (~30) is also reported (Fig. 2.6d). In recent years, such simple prevention measures were therefore increasingly replaced by soil or rock nailing (16 %), which had reduced effectively landslide

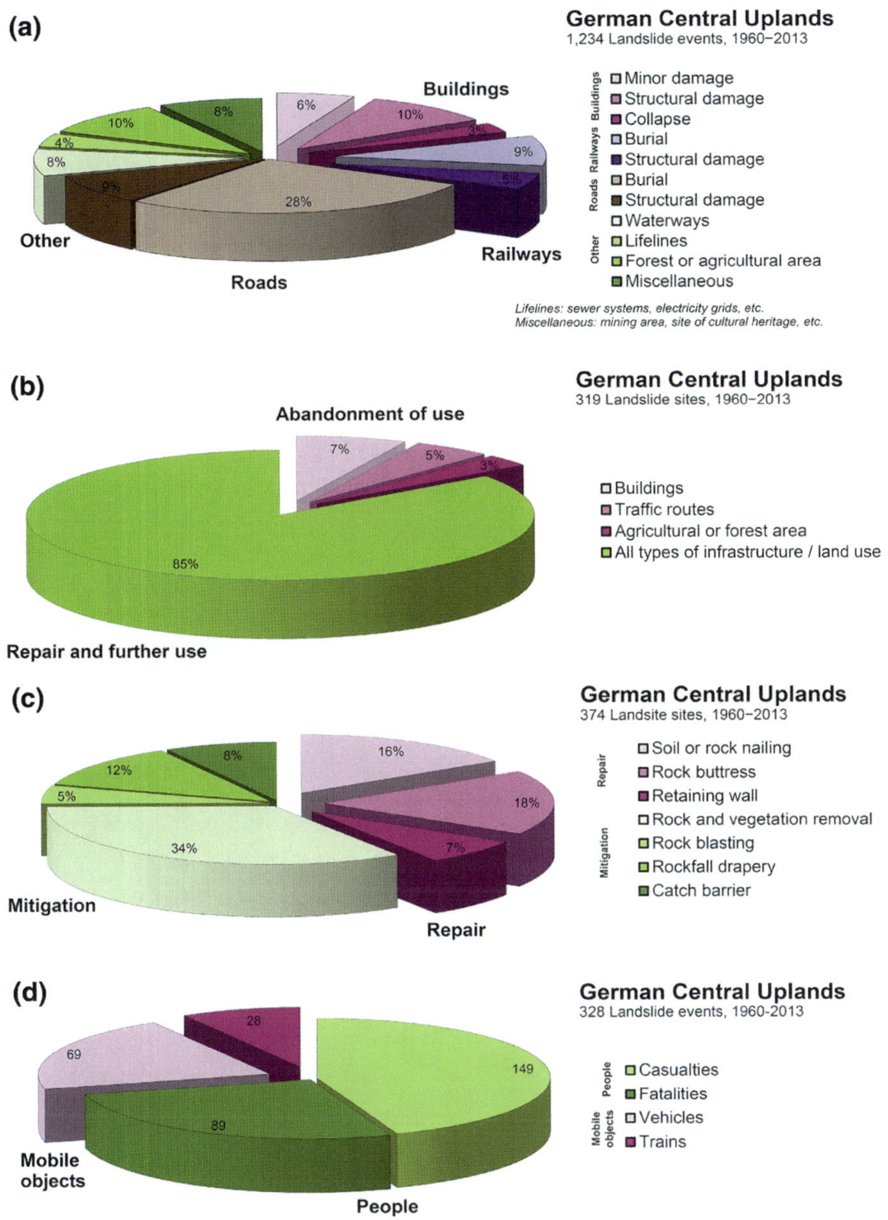

Fig. 2.6 Landslide impact and hazard mitigation in the German Central Uplands in the period 1960–2013. The figure shows the main types of affected infrastructure (**a**), prevailing forms of disaster response (**b**), and the most common methods of landslide repair and mitigation (**c**). Furthermore, an overview of the impact of landslides on people and mobile objects is given (**d**) (*Source* Damm and Klose 2015)

risk at many places (cf. Sect. 6.1). However, besides preventing landslide damage, its repair was required many times as well. A key role in repairing failed soil slopes, for example, was played by rock buttresses (18 %) and the use of retaining walls (7 %) (cf. Damm and Klose 2015).

References

Ackermann, E. (1959). Der Abtragungsmechanismus bei Massenverlagerungen an der Wellenkalkschichtstufe. *Zeitschrift für Geomorphologie NF, 3*, 193–226 und 283–304.

Barnikel, F., & Becht, M. (2004). Möglichkeiten einer Bewertung des regionalen Gefährdungspotentials auf der Basis historischer Naturgefahrenforschungen. *Zeitschrift für Geomorphologie NF, Suppl.-Bd. 135*, 1–10.

Battistini, A., Segoni, S., Manzo, G., Catani, F., & Casagli, N. (2013). Web data mining for automatic inventory of geohazards at national scale. *Applied Geography, 43*, 147–158.

Baum, I., & Schmidt, K.-H. (2001). Temporal classification of mass movements on the Wellenkalk-scarp in Thuringia and northern Hesse – Possibilities and limitations. *Zeitschrift für Geomorphologie NF, Suppl.-Bd. 125*, 25–41.

Baum, R. L., Highland, L. M., Lyttle, P. T., Fee, J. M., Martinez, E. M., & Wald, L. A. (2014). "Report a landslide" a website to engage the public in identifying geologic hazards. In K. Sassa, P. Canuti, & Y. Yin (Eds.), *Landslide Science for a Safer Geoenvironment, Vol. 1: The International Programme on landslides (IPL)* (pp. 95–100). Berlin: Springer.

Bibus, E., & Terhorst, B. (2001). Mass movements in Southwest Germany. Analyses and Results from the Tübingen Work Group of the MABIS Project. *Zeitschrift für Geomorphologie NF, Suppl.-Bd. 125*, 93–103.

Bock, B., Wehinger, A., & Krauter, E. (2012). Hanginstabilitäten in Rheinland-Pfalz: Ergebnisse der Rutschungsdatenbank Rheinland-Pfalz. *Mainzer Geowissenschaftliche Mitteilungen, 40*, 147–178.

Catani, F., Casagli, N., Ermini, L., Righini, G., & Menduni, G. (2005). Landslide hazard and risk mapping at catchment scale in the Arno River basin. *Landslides, 2*, 329–342.

CRED. (2014). EM-DAT: The OFDA/CRED International Disaster Database – www.emdat.be – Université catholique de Louvain – Brussels – Belgium. Retrieved July 10, 2014, from http://www.emdat.be/.

Creighton, R. (2006). The Irish landslides database. In R. Creighton (Ed.), *Landslides in Ireland* (pp. 9–22). Dublin: Geological Survey of Ireland, Irish Landslides Working Group.

Damm, B. (2005). *Gravitative Massenbewegungen in Mittelgebirgsräumen – Geowissenschaftlich-historische Analyse von Massenbewegungsprozessen in Nordhessen und Südniedersachsen als Beitrag zur Angewandten Geomorphologie* (262 S. + Anhang, unveröffentlicht). Habilitations schrift, Universität Eichstätt-Ingolstadt.

Damm, B., & Klose, M. (2014). Landslide database for the Federal Republic of Germany: A tool for analysis of mass movement processes and impacts. In K. Sassa, P. Canuti, & Y. Yin (Eds.), *Landslide Science for a Safer Geoenvironment, Vol. 2: Methods of landslide Studies* (pp. 787–792). Berlin: Springer.

Damm, B., & Klose, M. (2015). The landslide database for Germany: Closing the gap at national level. *Geomorphology*. doi:10.1016/j.geomorph.2015.03.021.

Damm, B., Becht, M., Varga, K., & Heckmann, T. (2010). Relevance of tectonic and structural parameters in Triassic bedrock formations to landslide susceptibility in Quaternary hillslope sediments. *Quaternary International, 222*, 143–153.

Devoli, G., Strauch, W., Chavez, G., & Høeg, K. (2007). A landslide database for Nicaragua: A tool for landslide-hazard management. *Landslides, 4*, 163–176.

Dikau, R., & Schmidt, K.-H. (2001). Mass movements in South, West and Central Germany – Objectives and main results of the MABIS project. *Zeitschrift für Geomorphologie NF, Suppl.-Bd. 125*, 1–12.

Dikau, R., Cavallin, A., & Jäger, S. (1996). Databases and GIS for landslide research in Europe. *Geomorphology, 15*, 227–239.

Foster, C., Pennington, C. V. L., Culshaw, M. G., & Lawrie, K. (2012). The national landslide database of Great Britain: Development, evolution and applications. *Environmental Earth Sciences, 66*, 941–953.

Galli, M., & Guzzetti, F. (2007). Landslide Vulnerability criteria: A case study from Umbria, central Italy. *Environmental Management, 40*, 649–664.

Galli, M., Ardizzone, F., Cardinali, M., Guzzetti, F., & Reichenbach, P. (2008). Comparing landslide inventory maps. *Geomorphology, 94*, 268–289.

Gans, P., Heinritz, G., Hemmer, I., Hemmer, M., Miener, K., Nienaber, B., et al. (2014). DGfG-Studie zum Image der Geographie – Anlage und erste Ergebnisse. *Rundbrief Geographie, 246*, 4–8.

Glade, T., & Crozier, M. (1996). Towards a national landslide information base for New Zealand. *New Zealand Geographer, 52*(1), 29–40.

Glade, T., Kadereit, A., & Dikau, R. (2001). Landslides at the Tertiary escarpments in Rheinhessen, Southwest Germany. *Zeitschrift für Geomorphologie NF, Suppl.-Bd. 125*, 65–92.

Gorum, T., Fan, X., van Westen, C. J., Huang, R. Q., Xu, Q., Tang, C., & Wang, G. (2011). Distribution pattern of earthquake-induced landslides triggered by the 12 May 2008 Wenchuan earthquake. *Geomorphology, 133*, 152–167.

Grunert, J., & Hardenbicker, U. (1991). Hangrutschungen im Bonner Raum – Ihre Genese und Kartierung für Planungszwecke. *Zeitschrift für Geomorphologie NF, Suppl.-Bd. 89*, 35–48.

Guzzetti, F., Reichenbach, P., Cardinali, M., Ardizzone, F., & Galli, M. (2003). The impact of landslides in the Umbria region, central Italy. *Natural Hazards and Earth System Sciences, 3*, 469–486.

Guzzetti, F., Ardizzone, F., Cardinali, M., Rossi, M., & Valigi, D. (2009). Landslide volumes and landslide mobilization rates in Umbria, central Italy. *Earth and Planetary Science Letters, 279*, 222–229.

Guzzetti, F., Mondini, A. C., Cardinali, M., Fiorucci, F., Santangelo, M., & Chang, K.-T. (2012). Landslide inventory maps: New tools for an old problem. *Earth-Science Reviews, 112*, 42–66.

Hardenbicker, U., & Grunert, J. (2001). Temporal occurrence of mass movements in the Bonn area. *Zeitschrift für Geomorphologie NF, Suppl.-Bd. 125*, 13–24.

Hervás, J. (2013). Landslide inventory. In P. T. Bobrowsky (Ed.), *Encyclopedia of Natural Hazards* (pp. 610–611). Berlin: Springer.

Hervás, J., & Bobrowsky, P. (2009). Mapping: Inventories, susceptibility, hazard and risk. In K. Sassa, & P. Canuti (Eds.), *Landslides – Disaster Risk Reduction* (pp. 321–349). Berlin: Springer.

Hilker, N., Badoux, A., & Hegg, C. (2009). The Swiss flood and landslide damage database 1972–2007. *Natural Hazards and Earth System Sciences, 9*, 913–925.

Hurst, M. D., Ellis, M. A., Royse, K. R., Lee, K. A., & Freeborough, K. (2013). Controls on the magnitude-frequency scaling of an inventory of secular landslides. *Earth Surface Dynamics, 1*, 67–78.

Jäger, D., Sandmeier, C., & Terhorst, B. (2012). Landslides and slope stability on the Franconian Alb, northern Bavaria. In *Extended Abstracts Interpraevent* (pp. 186–187). Grenoble, France.

Jaedicke, C., Lied, K., & Kronholm, K. (2009). Integrated database for rapid mass movements in Norway. *Natural Hazards and Earth System Sciences, 9*, 469–479.

Jelínek, R., Maitan, S., & Omura, H. (2001). Slope movements in Slovakia – Geographic and geological characteristics. *Journal of the Faculty of Agriculture, Kyushu University, 45*(2), 589–600.

Kirschbaum, D. B., Adler, R., Hong, Y., Hill, S., & Lerner-Lam, A. (2010). A global landslide catalog for hazard applications: Method, results, and limitations. *Natural Hazards, 52*, 561–575.

Klose, M., Damm, B., & Terhorst, B. (2014a). Landslide cost modeling for transportation infrastructures: A methodological approach. *Landslides*. doi:10.1007/s10346-014-0481-1.

Klose, M., Gruber, D., Damm, B., & Gerold, G. (2014b). Landslide susceptibility modeling on regional scales: The case of Lower Saxony, NW Germany. In K. Sassa, P. Canuti, & Y. Yin (Eds.), *Landslide Science for a Safer Geoenvironment, Vol. 2: Methods of Iandslide Studies* (pp. 437–442). Berlin: Springer.

Klose, M., Gruber, D., Damm, B., & Gerold, G. (2014c). Spatial databases and GIS as tools for regional landslide susceptibility modeling. *Zeitschrift für Geomorphologie NF, 58*(1), 1–36.

Klose, M., Highland, L., Damm, B., & Terhorst, B. (2014d). Estimation of direct landslide costs in industrialized countries: Challenges, concepts, and case study. In K. Sassa, P. Canuti, & Y. Yin (Eds.), *Landslide Science for a Safer Geoenvironment, Vol. 2: Methods of Iandslide Studies* (pp. 661–667). Berlin: Springer.

Komac, M., Fajfar, D., Ravnik, D., & Ribičič, M. (2007). Slovenian national landslide database – A promising approach to slope mass movement prevention plan. *Geologija, 50*, 393–402.

Kött, A., Grubert, A., & Aderhold, G. (2012). Die Hessische Rutschungs-Datenbank – Ein neues Instrument zur Erfassung, Archivierung und Auswertung von Massenbewegungen. In Hessisches Landesamt für Umwelt und Geologie (Hrsg.), *Jahresbericht 2012* (S. 89–102). Wiesbaden: Hessisches Landesamt für Umwelt und Geologie.

Krauter, E., Kumerics, C., Feuerbach, J., & Lauterbach, M. (2012). *Abschätzung der Risiken von Hang- und Böschungsrutschungen durch die Zunahme von Extremwetterereignissen* (61 S). Berichte der Bundesanstalt für Straßenwesen, Heft S75. Bremerhaven: Wirtschaftsverlag NW.

Liu, C., Li, W., Wu, H., Lu, P., Sang, K., Sun, W., et al. (2013). Susceptibility evaluation and mapping of China's landslides based on multi-source data. *Natural Hazards*. doi:10.1007/s11069-013-0759-y.

LKN-SH. (2014). Fachpläne Küstenschutz, Landesbetrieb für Küstenschutz, Nationalpark und Meeresschutz Schleswig-Holstein. Retrieved June 25, 2014, from http://www.schleswigholstein.de/KuestenSchutz/DE/kuestenSchutz_node.html.

Malamud, B. D., Turcotte, D. L., Guzzetti, F., & Reichenbach, P. (2004). Landslide inventories and their statistical properties. *Earth Surface Processes and Landforms, 29*, 687–711.

Mrozek, T., Kułak, M., Grabowski, D., & Wójcik, A. (2014). Landslide counteracting system (SOPO): Inventory database of landslides in Poland. In K. Sassa, P. Canuti, & Y. Yin (Eds.), *Landslide Science for a Safer Geoenvironment, Vol. 2: Methods of Iandslide Studies* (pp. 815–820). Berlin: Springer.

Munich Re. (2014). NatCatSERVICE. Retrieved July 15, 2014, from http://www.munichre.com/en/reinsurance/business/non-life/natcatservice/index.html.

Obst, K., & Schütze, K. (2010). The landslide database of Mecklenburg-Vorpommern and geohazard assessment at the Baltic Sea coast in NE Germany. In *Conference Proceedings Deuqua 2010* (pp. 58–59). Greifswald, Germany.

Osuchowski, M. (2008). The Australian landslide database. *AusGeo News, 92*, 17–19.

Pelletier, J. D., Malamud, B. D., Blodget, T., & Turcotte, D. L. (1997). Scale-invariance of soil moisture variability and its implications for the frequency-size distribution of landslides. *Engineering Geology, 48*, 255–268.

Petley, D. N. (2012). Global patterns of loss of life from landslides. *Geology, 40*, 927–930.

Petley, D. N., Dunning, S. A., & Rosser, N. J. (2005). The analysis of global landslide risk through the creation of a database of worldwide landslide fatalities. In O. Hungr, R. Fell, R. Couture, & E. Eberhardt (Eds.), *Landslide risk management* (pp. 367–373). London: A.A. Balkema Publishers, Taylor & Francis Group.

Rossi, M., Witt, A., Guzzetti, F., Malamud, B. D., & Peruccacci, S. (2010). Analysis of historical landslide time series in the Emilia-Romagna region, northern Italy. *Earth Surface Processes and Landforms, 35*(10), 1123–1137.

Schmidt, K.-H., & Beyer, I. (2001). Factors controlling mass movement susceptibility on the Wellenkalk-scarp in Hesse and Thuringia. *Zeitschrift für Geomorphologie NF, Suppl.-Bd. 125*, 43–63.

Schmidt, K.-H., & Beyer, I. (2003). High magnitude landslide-events on a limestone-scarp in central Germany – Morphometric characteristics and climatic controls. *Geomorphology, 49*, 323–342.

Schönwiese, C.-D. (2013). *Klimatologie*. Stuttgart: Ulmer. 489 S.

Schunke, E. (1971). Die Massenverlagerungen an den Schichtstufen und Schichtkämmen des Weser-Leine-Berglandes. *Nachrichten der Akademie der Wissenschaften in Göttingen, II. Mathematisch-Physikalische Klasse, 3*, 1–35.

Schweigl, J., & Hervás, J. (2009). *Landslide mapping in Austria* (61 pp). EC JRC-IES Report EUR 23785 EN. Luxembourg: Office for Official Publications of the European Communities.

Spizzichino, D., Margottini, C., Trigila, A., Iadanza, C., & Linser, S. (2010). Chapter 9: Landslides. In European Environment Agency (Ed.), *Mapping the impacts of natural hazards and technological accidents in Europe: An overview of the last decade* (pp. 81–93). EEA Technical report 13/2010. Copenhagen: European Environmental Agency.

Terhorst, B., & Kreja, R. (2009). Slope stability modelling with SINMAP in a settlement area of the Swabian Alb. *Landslides, 6*, 309–319.

Tonini, M., Pedrazzini, A., Penna, I., & Jaboyedoff, M. (2013). Spatial pattern of landslides in the Swiss Rhone Valley. *Natural Hazards*. doi:10.1007/s11069-012-0522-9.

Trigila, A., & Iadanza, C. (2008). *Landslides in Italy* (32 pp. + Appendix). Special Report 2008. Rome: Institute for Environmental Protection and Research.

Tsai, F., Hwang, J.-H., Chen, L.-C., & Lin, T.-H. (2010). Post-disaster assessment of landslides in southern Taiwan after 2009 Typhoon Morakot using remote sensing and spatial analysis. *Natural Hazards and Earth System Sciences, 10*, 2179–2190.

USGS. (2014). Landslide events. U.S. Geological Survey. Retrieved June 23, 2014, from http://landslides.usgs.gov/recent/.

Van Den Eeckhaut, M., & Hervás, J. (2012a). State of the art of national landslide databases in Europe and their potential for assessing landslide susceptibility, hazard and risk. *Geomorphology, 139–140*, 545–558.

Van Den Eeckhaut, M., & Hervás, J. (2012b). *Landslide inventories in Europe and policy recommendations for their interoperability and harmonization – A JRC contribution to the EU-FP7 SafeLand project* (202 pp). JRC Report EU 25666 EN. Luxembourg: Publications Office of the European Union.

Van Den Eeckhaut, M., Poesen, J., Vandekerckhove, L., Van Gils, M., & Van Rompaey, A. (2010). Human-environment interactions in residential areas susceptible to landsliding: The Flemish Ardennes case study. *Area, 42*(3), 339–358.

Varga, K., Becht, M., & Damm, B. (2006). Ansätze der GIS-gestützten räumlichen Modellierung von Rutschgefahren in Buntsandsteingebieten Nordhessens und Südniedersachsens (BRD). *Angewandte Geoinformatik, 2006*, 679–684.

Chapter 3
Study Area

3.1 Regional Setting

This research work dealing with landslide risk in NW Germany, Central Europe, is based on a regional and local study area in the Federal State of Lower Saxony, including: (i) the Lower Saxon Uplands, southern Lower Saxony, and (ii) the Upper Weser area, a 250-km^2 large area in the south of the Lower Saxon Uplands (Fig. 3.1). The entire region of southern Lower Saxony is located at the northern edge of the Central European Uplands and is rising sharply from the lowland areas of the North German Plain. This mountain region is characterized by low to moderate relief and elevations ranging from 50 to 950 m a.s.l. The total area of the Lower Saxon Uplands (core area) is about 7,400 km^2, which corresponds to 0.2 % of the total EU-28 territory. Three major physiographic areas can be differentiated in this region (e.g., Seedorf and Meyer 1992; Semmel 1996; Heunisch et al. 2007): (i) the Harz Mountains, a compact and strongly dissected Paleozoic basement complex, (ii) the Weser-Leine Uplands, where Mesozoic cuesta scarps and ridges rise abruptly above their forelands, and (iii) the Solling anticline, a gently undulating Triassic sandstone plateau with deeply incised river valleys.

Present-day topography in large parts of the Lower Saxon Uplands is mainly the result of late Jurassic to Tertiary geodynamic processes related to alpine tectonics (cf. Reicherter et al. 2008). This tectonic activity formed a complicated structural relief composed of rift or graben elements, complex fault and thrust fault structures, and a system of tilted, uplifted, and depressed blocks (e.g., Hedemann 1957; Lepper 1979; Wachendorf 1986; Drozdzewski 2003; Reicherter et al. 2008). In the Weser-Leine Uplands, for example, crustal fracturing and Tertiary erosion led to the evolution of a typical cuesta scarp landscape. Relief patterns in this area are characterized by a series of about 200 m high scarps and ridges that are built up of resistant Mesozoic lime- and sandstone (cf. Spönemann 1966, 1989; Schunke 1968a, b; Brunotte and Garleff 1980, 1989; Lehmeier 1988). Most of today's steep relief in areas such as the Solling anticline and the Harz Mountains (Fig. 3.1), by contrast, was formed by fluvial incision during the late Tertiary and Pleistocene. In combination with a strong uplift, especially of the Harz

M. Klose, *Landslide Databases as Tools for Integrated Assessment of Landslide Risk*, Springer Theses, DOI 10.1007/978-3-319-20403-1_3

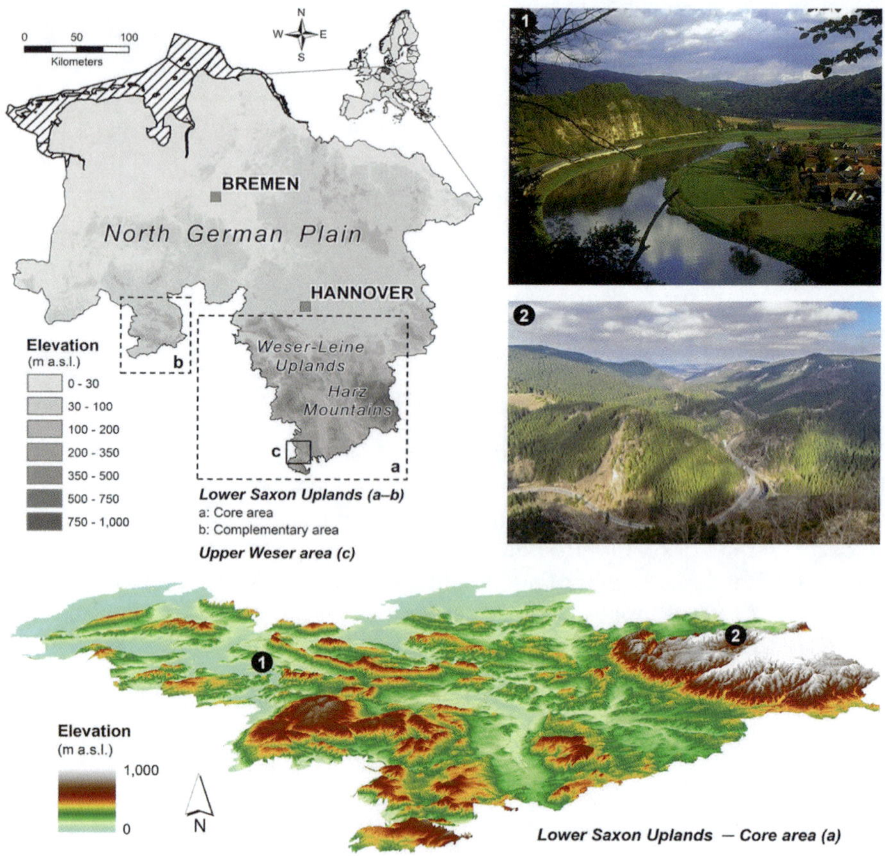

Fig. 3.1 Study areas in the Federal State of Lower Saxony, NW Germany. The regional focus of this work is on the Lower Saxon Uplands, including the Weser-Leine Uplands and the Harz Mountains as core area (**a**) and a complementary area (**b**) in the Osnabrücker Uplands. Both regions form together a 9,000-km^2 large mountain area in which elevations range between 50 and 950 m a.s.l. A local case study area (**c**) with a size of 250 km^2 is situated around the city of Hann. Münden in the Upper Weser area. Topographic relief in the Lower Saxon Uplands is often characterized by steep slopes: (*1*) Weser valley near the village of Dölme (Weser Uplands) with the Mühlenberg, a vertical limestone cut bank rising above highway B 83 (*Photo* Weserbergland Tourismus e.V.); and (*2*) Harz Mountains seen from the Ahrendsberger Klippen, a granite outcrop above the 300 m deep Oker valley that is located close to the city of Goslar (*Photo* D. Neumann), (ASTER GDEM, a product of METI and NASA)

Mountains, numerous erosion phases created up to 350 m deep gorges and river valleys (e.g., Brüning 1927; Hövermann 1950a, b; Mensching 1950; Amthauer 1972; Thiem 1972; Rohde 1989). The dominating river valley system is that of the Weser river and its main tributaries (Fulda river, Werra river; see also Grupe 1926), which is eroded into the Bunter Sandstone formations of the Solling anticline (e.g., Hedemann 1957; Backhaus et al. 1958, 1980).

Except for the Harz Mountains, elevations rarely exceed 500 m a.s.l., but still relief intensity is comparatively high in this region, especially at the main scarps that usually show slope gradients of up to 40° (e.g., Schunke 1968a). Highest values of terrain steepness, however, most often exist in river valleys, where slope gradients >45° are typical at cut banks or in canyons and gorges (cf. Brüning 1927; Garleff 1985). As result of tectonic stress and intensive Tertiary weathering, bedrock in the Lower Saxon Uplands, most notably in the Upper Weser area, is strongly disintegrated, thus comparatively weak (cf. Damm 2005a; Müller 2009; Damm et al. 2010). Bedrock throughout the entire region is covered to large extent by Quaternary sediments of several meter thickness (e.g., Hövermann 1953; Suchel 1954; Schilling and Wiefel 1962; Frühauf 1991; Wagner 2011). The topographic setting of the Lower Saxon Uplands together with the properties of soil and bedrock is from a regional perspective a key factor of the widespread landslide susceptibility in this mountain area.

3.2 Landslide Types, Processes, and Materials

Landslides in the Lower Saxon Uplands occur on annual basis in a variety of different types and magnitudes (Fig. 3.2). According to the available landslide database subset for Lower Saxony (Sect. 4.2.2) and the existing literature (see below), many landslides in this region are characterized by complex types of movement (Fig. 3.2c, f), with different states or styles of activity and variable movement rates. Much of these complex landslides represent large slope movements along the scarps and ridges of the Weser-Leine Uplands, where soft beds of marl- and claystone are overlain by permeable lime- and sandstone strata (e.g., Heunisch et al. 2007). Such complex landslides are commonly classified as rock topple–earth flow, with rock spreading above liquefied substratum characterizing movement mechanisms at the crown of many slides. Further downslope, toppling of loose rock from the main scarp or displaced spreading blocks is frequently observable, and in the foot zone or at the slide toe, slow creep or flow processes in weathered debris or soil material dominate landslide movement. The ages of these complex landslides are highly variable, including slide masses of late Pleistocene age (relict landslides), several generations of Holocene (ancient) or pre-modern slope movements, and few recent landslides of such large spatial extent. Many of these landslides are at least in part active today or are periodically reactivated during wet periods or because of construction works (cf. Ackermann 1953; Mortensen and Hövermann 1956; Mortensen 1960; Schunke 1971; Stein 1975; Tilch 1999; Bense et al. 2011). For further information, see also related studies from adjacent areas in northern Hesse or NW Thuringia, for example, Ackermann (1958a, b, 1959), Bernhard (1967), Krümmling et al. (1975), and Rösing and Wenzel (1989).

The most common landslide types over the past decades, however, have been shallow translational or rotational slides in soil material (Fig. 3.2d). As often occurring in highly saturated soils, these slide processes frequently convert to flow–slides, wherefore often referred to as slump–earth flows. Further main landslide types in the Lower Saxon Uplands are small or medium large rockfalls, block

Fig. 3.2 (continued)

Fig. 3.2 (continued)

Fig. 3.2 (continued)

◀ **Fig. 3.2** Examples of landslides in the Lower Saxon Uplands, NW Germany: **a** Mitigation of a year-1995 planar rockslide in Lower Carboniferous greywacke along highway B 498 at the Oker reservoir (Harz Mountains) near the city of Goslar (*Photo* Database B. Damm). **b** Deep-seated rotational slide as result of excavating the Mittelland Canal in the early 20th century. The photo dates back to 1928 and was probably taken outside the city of Bramsche (*Photo* State Archive Hannover, HSTAH Bigs.Nr.5553-2). **c** The 2004 Messingsberg rockslide–rock avalanche with a volume of 500,000 m^3 and with up to 30 m deep tension cracks caused by mining activity at an Upper Jurassic limestone ridge in the Weser Mountains (*Photos* B. Scheel; T. Landmann; see also Meyer 2005; NNG 2013). **d** A May 2013 rainfall-induced translational slide in slope deposits and Bunter Sandstone bedrock at a steep road cut in the city of Hann. Münden (*Photo* P. Maurischat). **e** A 20,000-m^3 large debris flow affected a neighborhood of the city of Osnabrück after a dam failure of a settling pond during a period of heavy rainfall in the year 1957 (*Photo* Archiv Museum Industriekultur; see also THW 1957). **f** Complex landslide located at an Upper Jurassic limestone ridge near the town of Brunkensen. The 4 ha large landslide was reactivated in 1988 after a period of rapid snowmelt and heavy rainfall (*Photos* N. Tilch; D. Garbermann; see also Tilch 1999)

toppling, and planar rockslides (Fig. 3.2a). These types of movement are widespread along steep cut banks of the Weser river or at eroded rock cliffs in the many steep Harz Mountain valleys. Above landslide types typically involve both natural and man-made slopes, although ones affecting made-made slopes (i.e., cut or fill slopes; Fig. 3.2b, d, e) were more frequent in the recent past, especially along the traffic routes of this region (e.g., Günther et al. 2004, 2012; Damm 2005a; Welsch 2006; Damm et al. 2010; Gidde 2012; Klose et al. 2012).

The different types of landslides in the Lower Saxon Uplands vary moderately in size, with a majority of them being classified as small landslides. This applies especially to shallow soil slides in thin Quaternary cover beds as well as much of the recorded rockfalls, both of which rarely exceed volumes of >300–500 m^3. As database analysis has shown, landslides with volumes up to several few thousand cubic meters are located either in slope toe positions or on large cut and fill slopes (Fig. 3.2b, d), whereby these slope failures most often represent rotational slides (see also Damm 2005b; Damm et al. 2010). Besides slope creep in deeply weathered bedrock, the above complex landslide types are among the largest mass movements in this mountain area. Landslide volumes in these cases can reach half a million cubic meters or more (Fig. 3.2c), while the displaced material of such landslides sometimes extends over 5–15 ha (e.g., Ackermann 1953; Tilch 1999; Meyer 2005; Fig. 3.2f). The largest recorded slope movement by volume in recent times is with 600,000–800,000 m^3 the 1961 Rattberg landslide in the Upper Weser area which covers an area of about 16 ha (cf. Kleine-Möllhoff 2003; Damm 2005b). Landslide velocity usually varies from very slow (mm/year) to extremely rapid (m/s) movement, and movement rates are not only varying between the different landslide types, but also within many complex landslides themselves. Furthermore, the spatial distribution of activity in complex landslides is highly variable, including dormant or slow-moving active landslide parts and areas where rapid reactivation of slide masses takes place (cf. Schunke 1971; Tilch 1999; Damm 2005a; Meyer 2005). According to the regional size and velocity distribution of landslides, landslide magnitude in the Lower Saxon Uplands is considered

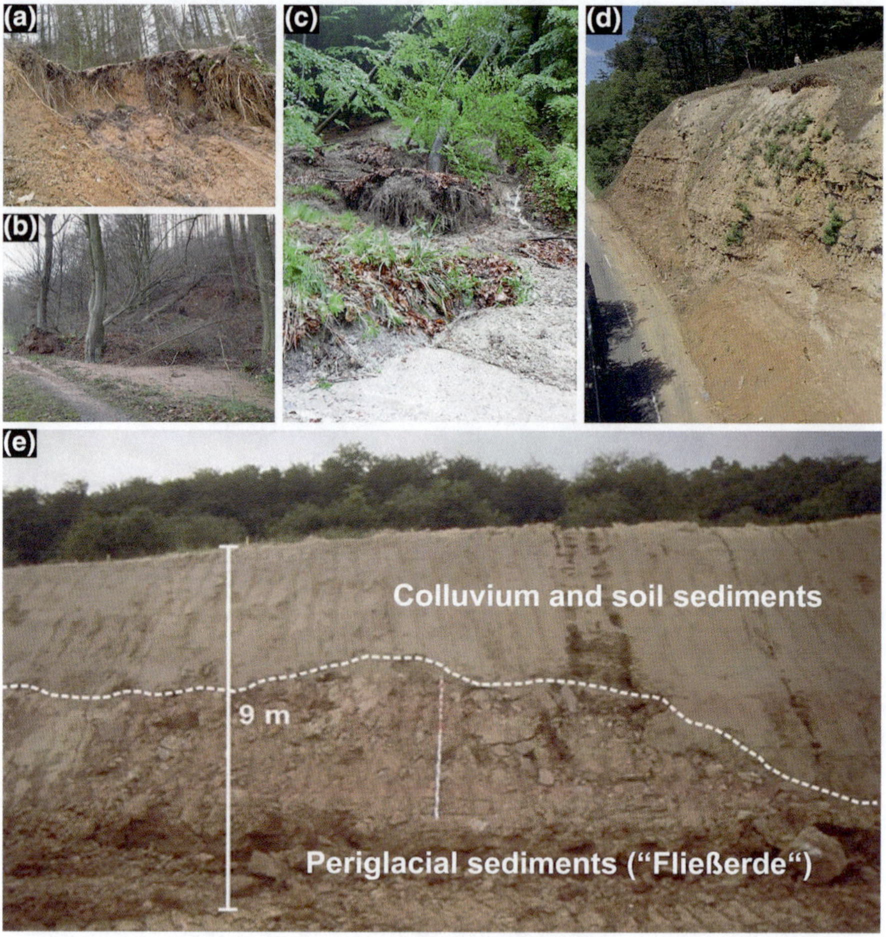

Fig. 3.3 Landslides in Quaternary cover beds in the Lower Saxon Uplands and vertical structure of a typical cover bed complex in the Upper Weser area: **a** Scarp of a shallow landslide in 2 m thick periglacial sediments that overlie Bunter Sandstone bedrock. The year-2009 landslide buried parts of highway B 80 north of the town of Reinhardshagen (*Photo* Database B. Damm). **b** Failure of the B 80 highway embankment constructed of soil material that originates from Pleistocene slope deposits. Heavy rainfall caused this landslide near Reinhardshagen in November 2010 (*Photo* M. Klose). **c** Shallow landslide in highly saturated soil triggered by the May 2013 rainstorms in the Deister mountain area southeast of the town of Lauenau (*Photo* Niedersächsische Landesforsten). **d** Cover beds susceptible to landslides overlying a steep outcrop of disintegrated sandstone strata (smH) of the Bunter Sandstone formation. The landslide site is situated above highway L 561 north of the city of Hann. Münden (*Photo* Database B. Damm). **e** Profile showing the vertical structure of the regionally widespread cover bed complex for a slope toe position at a landslide site in the Schede valley near Hann. Münden (*Photo* Database B. Damm)

to be small to moderate, since large slope movements rarely show high velocities, thus extreme kinetic energy (landslide intensity).

Much of the landslide activity in the Lower Saxon Uplands is taking place in widespread Quaternary cover beds that overlie bedrock on many slopes (e.g., Damm et al. 2010, 2013; Klose et al. 2012; Fig. 3.3). With regard to the Upper Weser area, near-surface subsoil is characterized by a three-part Quaternary cover bed complex (see below, cf. Emmerich 1997), and underlying bedrock in this area includes Triassic rock formations of the Middle Bunter Sandstone (sm). The sm mostly consists of laminated to stratified fine- to medium-grained sandstones with interbeds of fragile silt- or claystones (cf. Lohmann 1959; Neumann-Redlin and Lepper 1975; Backhaus et al. 1958, 1980). Tectonic movement and deep Tertiary weathering caused a strong disintegration and decomposition of solid rock (e.g., Hedemann 1957; Müller 2009), which is not only a key factor for slope instability in bedrock, but also in overlying cover beds, especially as result of leaking strata and fissure water (cf. Damm 2005a; Damm et al. 2010; Fig. 3.3d). The thickness of cover beds is about 2–4 m on average, yet varies in relation to local topography and weathering intensity, with thicknesses of 15 m and more often existing at the toe of slopes (Fig. 3.3a, e; see also Thomas 1993). In the Upper Weser area, the following three-part cover bed complex is frequently observed (cf. Emmerich 1997; Damm et al. 2010, 2013; Klose et al. 2012):

Basal rock debris. The transition zone between bedrock and subsoil is marked by a several decimeters to a few decameters thick weathering layer that is to a large extent composed of rock debris. This layer is mainly the result of deep bedrock weathering during the Tertiary and downslope deposition of weathering products. Coarse soil (50–80 %) consists of angular debris of the cobble and large boulder fraction (>630 mm), while grain sizes in fine soil show a maximum in the range of sandy loams and loamy sands. Loess content is low or lacking, and the transition to the overlying periglacial sediments is usually diffuse (cf. Damm et al. 2013). Note that for the English reader the term "basal rock debris" better describes the lowermost unit of the cover bed complex than "mantle rock" as it is referred to in Sect. 5.2.

Periglacial sediments ("Fließerde"). This up to 3 m thick layer of solifluction debris is primarily the product of periglacial slope processes during the Pleistocene. The soil fabric is characterized by a mixture of coarse to fine-grained material, with coarse soil (30–60 %) widely consisting of angular debris of the cobble and the boulder fraction (≥150–200 mm). In the fine soil, the grain size maximum lies between sandy to clayey silts or silty to loamy sands, and most field samples usually show a coherent fabric. Eolian material is widespread in this sediment complex but is often found in variable proportions. Dried samples show soil cracks with average widths of 5–20 mm that facilitate water infiltration in case of moisture supply. In normal field conditions, periglacial sediments are moderately to strongly consolidated, thus effective compactness is of intermediate values. For further information on soil physical and soil mechanical properties, it is referred to Sect. 5.2.1 (cf. Damm et al. 2013; see also Thomas 1983; Fig. 3.3e).

Colluvium and soil sediments. The uppermost unit of the cover bed complex is represented by colluvial sediments of eroded loesses or loess loams and/or a soil formation on top of the underlying sediments. Colluvium dominates especially

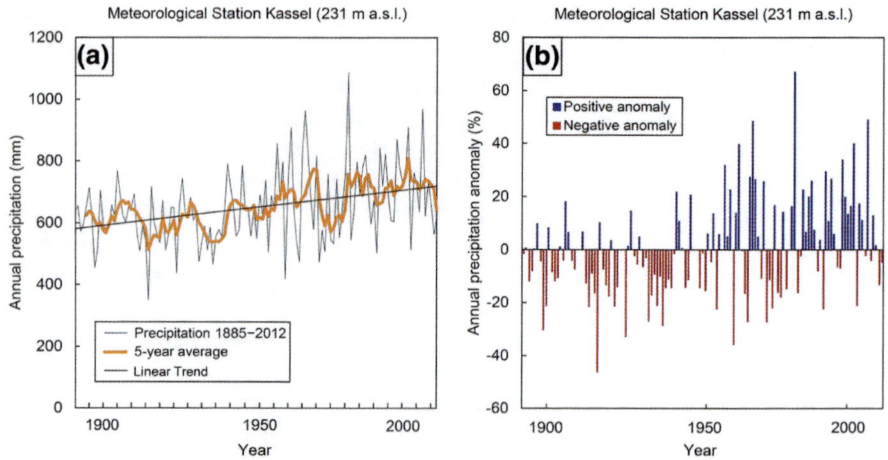

Fig. 3.4 Annual rainfall trend (**a**) and anomaly (**b**) for the meteorological station of Kassel (northern Hesse) between 1885 and 2012. This reference station is located close to the Upper Weser area and approximates the regional rainfall conditions with some critical limitations (cf. Sect. 4.1.2). The illustrated rainfall trend shows a linear increase in annual precipitation of 24 % and is characterized by high annual variability. Significant dry or wet periods over the past 130 years had been the periods 1907–1911, 1913–1918, and 1927–1938 or 1965–1968, 1979–1981, and 1998–2002. The figures show an increase in the number of years with above-average humidity (wet years) over time as well as an increase in the total precipitation of such wet years (*Data source* German Meteorological Service, http://www.dwd.de/)

at the toe of slopes where its transition to beneath periglacial sediments is often marked by a sharp boundary. This layer of soil sediments, as it is referred to in Sect. 5.2.1, shows thicknesses varying between several decimeters on slopes, with the tendency to increase downslope to a several meter thick sediment complex. In slope toe positions, for instance, average sediment thickness is about 3–6 m. The maximum in the grain size distribution of this layer is in the range of coarse silt (cf. Damm et al. 2013; see also Bork 1985; Fig. 3.3e).

Landslides in hillslope sediments of this region often occur during prolonged wet periods when soils are highly saturated (see Sect. 5.2; Fig. 3.3c). Using the example of the meteorological station Kassel (cf. Sect. 4.1.2), climate statistics demonstrates a linear increase of rainfall of 24 % since the year 1885 (Fig. 3.4a). The rainfall trend for this reference station close to the Upper Weser area is characterized by a high temporal variability over the past 130 years, which manifests in a strong fluctuation of periods with above- and below-average humidity. There are thus strong anomalies in the annual amount of rainfall (Fig. 3.4b), with variations from the 1885–2012 annual mean (650 mm) ranging between +67 % (year 1981, 1087 mm) and −46 % (year 1911, 350 mm). Since the late 19th century, there has not only been a shift to an increased number of years or months with above-average humidity, but also, and more importantly, total precipitation in such extreme years or months has significantly increased as well. This is accompanied by a recent rainfall trend that suggests a significant increase (~15 %) in winter precipitation since the mid-1970s (Fig. 3.4b; cf. Damm 2005b).

References

Ackermann, E. (1953). Der aktive Bergrutsch südlich der Mackenröder Spitze in geologischer Sicht. Nachrichten der Akademie der Wissenschaften in Göttingen, Mathematisch-Physikalische Klasse, IIa. *Mathematisch-Physikalisch-Chemische Abteilung, 5*, 67–83.

Ackermann, E. (1958a). Die Sturzfließung am Schickeberg südlich Eschwege. *Notizblatt Hessisches Landesamt für Bodenforschung zu Wiesbaden, 87*, 172–187.

Ackermann, E. (1958b). Bergstürze und Schuttströme an der Wellenkalk-Schichtstufe Mitteldeutschlands in Gegenwart und Vergangenheit. *Natur und Volk, 88*(4), 123–132.

Ackermann, E. (1959). Der Abtragungsmechanismus bei Massenverlagerungen an der Wellenkalkschichtstufe. *Zeitschrift für Geomorphologie NF, 3*, 193–226 und 283–304.

Amthauer, H. (1972). Untersuchungen zur Talgeschichte der Oberweser. *Göttinger Geographische Abhandlungen, 59*, 99 S.

Backhaus, E., Lohmann, H., & Regenhardt, H. (1958). Der Mittlere Buntsandstein im Reinhardswald (Nordhessen). *Notizblatt des hessischen Landesamtes für Bodenforschung, 86*, 192–201.

Backhaus, E., Gramann, F., Kaever, M., Lepper, J., Lohmann, H. H., Meiburg, P., et al. (1980). *Erläuterungen zur Geologischen Karte des Reinhardswaldes 1:50.000*. Wiesbaden: Hessisches Landesamt für Bodenforschung. 32 S.

Bense, F. A., Ertl, G., Vollbrecht, A., & Battaglia, M. (2011). Gravitative Massenverlagerungen an der Röt/Muschelkalk-Schichtstufe des Göttinger Waldes – GPS-gestützte Strukturkartierungen. In B. Leiss, D. Tanner, A. Vollbrecht, & G. Arp (Hrsg.), *Neue Untersuchungen zur Geologie der Leinetalgrabenstruktur: Bausteine zur Erkundung des geothermischen Potentials der Region Göttingen* (pp. 125–131). Göttingen: Universitätsverlag Göttingen.

Bernhard, H. (1967). Massenverlagerungen an der Röt-Muschelkalk-Grenze. *Zeitschrift der Deutschen Geologischen Gesellschaft, 119*, 585–586.

Bork, H.-R. (1985). Mittelalterliche und neuzeitliche lineare Bodenerosion in Südniedersachsen. *Hercynia N.F., 22*, 259–279.

Brunotte, E., & Garleff, K. (1980). Tectonic and climatic factors of landform development on the northern fringe of the German Hill Country (Deutsche Mittelgebirge) since the early Tertiary. *Zeitschrift für Geomorphologie NF, Suppl.-Bd. 36*, 104–112.

Brunotte, E., & Garleff, K. (1989). Structural landforms and planation surfaces in southern Lower Saxony. *Catena Supplement, 15*, 151–164.

Brüning, K. (1927). Die Reliefenergie des Harzes. In K. Brüning (Hrsg.), *Jahrbuch der Geographischen Gesellschaft zu Hannover für das Jahr 1927* (S. 44–50). Hannover.

Damm, B. (2005a). Gravitative Massenbewegungen in Südniedersachsen. Die Altmündener Wand – Analyse und Bewertung eines Rutschungsstandortes. *Zeitschrift für Geomorphologie NF, Suppl.-Bd. 138*, 189–209.

Damm, B. (2005b). *Gravitative Massenbewegungen in Mittelgebirgsräumen – Geowissenschaftlich-historische Analyse von Massenbewegungsprozessen in Nordhessen und Südniedersachsen als Beitrag zur Angewandten Geomorphologie* (262 S. + Anhang, unveröffentlicht). Habilitationsschrift, Universität Eichstätt-Ingolstadt.

Damm, B., Becht, M., Varga, K., & Heckmann, T. (2010). Relevance of tectonic and structural parameters in Triassic bedrock formations to landslide susceptibility in Quaternary hillslope sediments. *Quaternary International, 222*, 143–153.

Damm, B., Terhorst, B., & Ottner, F. (2013). Geotechnical properties of cover beds. In A. Kleber, & B. Terhorst (Eds.), Mid-latitude slope deposits (cover beds). *Developments in Sedimentology, 66*, 153–170.

Drozdzewski, G. (2003). Geologische Entwicklung und tektonischer Bau. In H.-J. Betzer, W. Dassel, G. Drozdzewski, J. Farrenschon, A. Gawlik, H. Heuser, D. Juch, K.-H. Ribbert, & K. Skupin (Eds.), *Geologie im Weser- und Osnabrücker Bergland* (pp. 16–30). Krefeld: Geologischer Dienst Nordrhein-Westfalen.

Emmerich, K. H. (1997). *Erläuterungen zur Bodenkarte von Hessen 1:25.000. Blatt Nr. 4523 Münden*. Wiesbaden: Hessisches Landesamt für Bodenforschung. 113 S.

Frühauf, M. (1991). Neue Befunde zur Lithologie, Gliederung und Genese der periglazialen Lockermaterialdecken im Harz: Fremdmaterialnachweis und Decksedimenterfassung. *Petermanns Geographische Mitteilungen, 135*(4), 249–255.

Garleff, K. (1985). *Erläuterungen zur Geomorphologischen Karte der Bundesrepublik Deutschland, Blatt 2 C 4722 Kassel*. Berlin. 74 S.

Gidde, A. (2012). Niederschlagsabhängige Böschungsschäden in Niedersachsen. In Bundesanstalt für Straßenwesen (Hrsg.), *42. Erfahrungsaustausch über Erdarbeiten im Straßenbau* (S. 90–92). Berichte der Bundesanstalt für Straßenwesen, Heft S 76. Bremerhaven: Wirtschaftsverlag NW.

Grupe, O. (1926). Tal- und Terrassenbildung im Gebiete der Werra-Fulda-Weser und Soergels „Gliederung und absolute Zeitrechnung des Eiszeitalters". *Geologische Rundschau, 17*(3), 161–196.

Günther, A., Carstensen, A., & Pohl, W. (2004). Automated sliding susceptibility mapping of rock slopes. *Natural Hazards and Earth System Sciences, 4*, 95–102.

Günther, A., Wienhöfer, J., & Konietzky, H. (2012). Automated mapping of rock slope geometry, kinematics and stability with RSS-GIS. *Natural Hazards, 61*, 29–49.

Hedemann, H.-A. (1957). Die Gewölbestruktur des Sollings und ihre Entstehung. *Geologisches Jahrbuch, 72*, 529–638.

Heunisch, C., Caspers, G., Elbracht, J., Langer, A., Röhling, H.-G., Schwarz, C., & Streif, H. (2007). Erdgeschichte von Niedersachsen. Geologie und Landschaftsentwicklung. *GeoBerichte, 6*, 3–85.

Hövermann, J. (1950a). Die diluvialen Terrassen des Oberharzes und seines Vorlandes. Ein Beitrag zur Frage der Harzhebung. *Petermanns Geographische Mitteilungen, 94*, 121–130.

Hövermann, J. (1950b). Die Oberflächenformen des Harzes. *Geographische Rundschau, 2*, 208–212.

Hövermann, J. (1953). Die Periglazial-Erscheinungen im Harz. In H. Poser (Hrsg.), Studien über die Periglazial-Erscheinungen in Mitteleuropa. Teil I, Studien aus Mittelgebirgen (Harz, Schwarzwald, Werratal). *Göttinger Geographische Abhandlungen, 14*, S. 7–44.

Kleine-Möllhoff, K. (2003). *Ursachen und Auslöser von Hangrutschungen in Buntsandsteingebieten – Dargestellt an Beispielen aus Nordhessen* (109 S. + Anhang). Diplomarbeit, Georg-August-Universität Göttingen, Geographisches Institut.

Klose, M., Damm, B., & Gerold, G. (2012). Analysis of Landslide Activity and Soil Moisture in Hillslope Sediments using a Landslide Database and a Soil Water Balance Model. *GEO-ÖKO, 33*(3–4), 204–231.

Krümmling, H., Tornack, E., Wiefel, J., & Wucher, K. (1975). Massenverlagerungen an der Röt-Muschelkalk-Schichtstufe Nordwest-Thüringens. *Zeitschrift für Angewandte Geologie, 21*, 552–558.

Lehmeier, F. (1988). Zum Formenschatz der Schichtkammlandschaft im Niedersächsischen Bergland (insbesondere im Ith-Hils-Bergland). *Berliner Geographische Abhandlungen, 47*, 49–61.

Lepper, J. (1979). Zur Struktur des Solling-Gewölbes. *Geologisches Jahrbuch A, 51*, 57–77.

Lohmann, H. (1959). *Zum Bau des Oberweserberglandes zwischen Hannoversch-Münden und Karlshafen* (233 S). Dissertation, Universität Hamburg, Mathematisch-Naturwissenschaftliche Fakultät.

Mensching, H. (1950). Schotterfluren und Talauen im Niedersächsischen Bergland. *Göttinger Geographische Abhandlungen, 4*, 54 S.

Meyer, S. (2005). Der rezente Bergsturz am Messingsberg im niedersächsischen Wesergebirge. *Mitteilungen des Verbandes der Deutschen Höhlen- und Karstforscher, 51*(3), 96–100.

Mortensen, H. (1960). Neues über den Bergrutsch südlich der Mackenröder Spitze und über die holozäne Hangformung an Schichtstufen im mitteleuropäischen Klimabereich. *Zeitschrift für Geomorphologie NF, Suppl.-Bd. 1*, 114–123.

Mortensen, H., & Hövermann, J. (1956). Der Bergrutsch an der Mackenröder Spitze bei Göttingen (pp. 149–155). In *Premier Rapport de la Commission pour l'Étude des Versants*. Préparé pour le Congrès International de Géographie, Rio de Janeiro, 1956.

Müller, C. (2009). *Charakterisierung des hydromechanischen Verhaltens der Gesteine des Mittleren Buntsandsteins im Hinblick auf eine geothermische Nutzung: Strukturgeologische Geländeaufnahmen, gesteinsmechanische Untersuchungen und numerische Modellierungen* (155 S). Dissertation, Georg-August-Universität Göttingen, Mathematisch-Naturwissenschaftlichen Fakultäten. Retrieved September 14, 2014, from https://ediss.uni-goettingen.de/handle/11858/00-1735-0000-0006-B2E2-B?show=full.

Neumann-Redlin, C., & Lepper, J. (1975). Profile im Mittleren Buntsandstein des Bramwaldes (Südniedersachsen). *Geologisches Jahrbuch A, 29*, 57–64.

NNG. (2013). Bergrutsch Steinbergen. Norddeutsche Naturstein GmbH. Retrieved July 26, 2014, from http://www.bergrutsch-steinbergen.de.

Reicherter, K., Froitzheim, N., Jarsiński, M., Badura, J., Franzke, H.-J., Hansen, M., et al. (2008). Alpine tectonics north of the Alps. In T. McCann (Ed.), *The Geology of Central Europe, Vol. 2: Mesozoic and Cenozoic* (pp. 1233–1285). London: Geological Society.

Rohde, P. (1989). Elf pleistozäne Sand-Kies-Terrassen der Weser: Erläuterung eines Gliederungsschemas für das obere Weser-Tal. *Eiszeitalter und Gegenwart, 39*(1), 42–56.

Rösing, F., & Wenzel, B. (1989). Der Bergrutsch am Nordhang der Hörne bei Bad Sooden-Allendorf (Nordhessen). *Geologisches Jahrbuch Hessen, 117*, 237–250.

Schilling, W., & Wiefel, H. (1962). Jungpleistozäne Periglazialbildungen und ihre regionale Differenzierung in einigen Teilen Thüringens und des Harzes. *Geologie, 11*(4), 428–460.

Schunke, E. (1968a). Die Schichtstufenhänge im Weser-Leine-Bergland in Abhängigkeit vom geologischen Bau und Klima. *Göttinger Geographische Abhandlungen, 43*, 219 S.

Schunke, E. (1968b). Die Schichtstufenhänge des Leine-Weser-Berglandes – Methoden und Ergebnisse ihrer Untersuchung. *Geologische Rundschau, 58*(2), 446–464.

Schunke, E. (1971). Die Massenverlagerungen an den Schichtstufen und Schichtkämmen des Weser-Leine-Berglandes. *Nachrichten der Akademie der Wissenschaften in Göttingen, II. Mathematisch-Physikalische Klasse, 3*, 1–35.

Seedorf, H. H., & Meyer, H.-H. (1992). *Landeskunde Niedersachsen – Natur- und Kulturgeschichte eines Bundeslandes*. Band 1: Historische Grundlagen und naturräumliche Ausstattung. Neumünster: Wachholtz. 517 S.

Semmel, A. (1996). *Geomorphologie der Bundesrepublik Deutschland. Grundzüge, Forschungsstand, aktuelle Fragen, erörtert an ausgewählten Landschaften*. Erdkundliches Wissen, Heft 30. Stuttgart: Franz Steiner Verlag. 199 S.

Spönemann, J. (1966). Geomorphologische Untersuchungen an Schichtkämmen des Niedersächsischen Berglandes. *Göttinger Geographische Abhandlungen, 36*, 167 S.

Spönemann, J. (1989). Homoclinal ridges in Lower Saxony. *Catena Supplement, 15*, 133–149.

Stein, C. (1975). Studien zur quartären Talbildung in Kalk- und Sandgesteinen des Weser-Leine-Berglandes. *Göttinger Geographische Abhandlungen, 64*, 131 S.

Suchel, A. (1954). Studien zur quartären Morphologie des Hilsgebietes. In H. Poser (Ed.), Studien über die Periglazial-Erscheinungen in Mitteleuropa, Teil IV. *Göttinger Geographische Abhandlungen, 17*, 147 S.

Thiem, W. (1972). Geomorphologie des westlichen Harzrandes und seiner Fußregion. *Jahrbuch der Geographischen Gesellschaft zu Hannover, Sonderheft 6*, 271 S.

Thomas, J. (1993). Untersuchungen zur holozänen fluvial Geomorphodynamik an der oberen Oberweser. *Göttinger Geographische Abhandlungen, 98*, 111 S.

THW (1957) (Autor unbekannt). Dammbruch mit schweren Folgen. *Das Technische Hilfswerk, 4*(12), 7–8.

Tilch, N. (1999). Rutschungs-Suszeptibilität im südlichen Niedersachsen – Von der Anatomie der Rutschung bei Brunkensen/Alfeld zur Prognose instabiler Hanglagen. *Braunschweiger Geowissenschaftliche Arbeiten, 22*, 184 S.

Wachendorf, H. (1986). Der Harz – Variszischer Bau und geodynamische Entwicklung. *Geologisches Jahrbuch A, 91*, 3–67.

Wagner, B. (2011). Spatial analysis of loess and loess-like sediments in the Weser-Aller catchment (Lower Saxony and Northern Hesse, NW Germany). *E&G Quaternary Science Journal, 60*(1), 27–46.

Welsch, M. (2006). *Die Risikoanalyse als Verfahren zur Beschreibung von Gefahrenpotenzialen und Risiken durch gravitative Massenbewegungen in Mittelgebirgsräumen. Am Beispiel von ausgewählten Strassenabschnitten im Oberwesergebiet (Niedersachsen)* (76 S). Diplomarbeit, Georg-August-Universität Göttingen, Geographisches Institut.

Chapter 4
Methodology

4.1 Soil Water Balance Model

4.1.1 Introduction

Landslides in Central Europe are geomorphic processes that are primarily controlled by soil hydrology (e.g., Van Beurden 1997; Theisen 1998; Boogard and Van Asch 2002; Prokešová et al. 2013). Soil moisture conditions, which are mainly regulated by precipitation, snowmelt and ground frost, play therefore a major role in landslide initiation (e.g., Van Asch et al. 1999; Sidle and Ochiai 2006; Lu and Godt 2013), especially in temperate regions such as Central Europe. Changes in water content and the level of soil saturation have direct influence on the shear strength of soils and are thus closely connected with slope instability (e.g., Iverson 2000; Cho and Lee 2001; Tsai and Chen 2010). The effect of water influx on soil strength is dependent on the grain sizes of soils, their permeability and plasticity as well as the main soil-related shear parameters, including weight, internal friction, and cohesion (e.g., Knoblich 1967; Duncan and Wright 2005; De Blasio 2011). Water infiltration into hillslope sediments causes a reduction of shear strength by decreasing cohesion and by leading to an increase in pore water pressure and hydrostatic pressure (see also Lu and Likos 2004). These soil hydrological processes show the potential to trigger to landslides or to reduce the stability of slopes to a marginally stable level (e.g., Crozier 1986; Collins and Znidarcic 2004; Godt et al. 2009, 2012). Given the mechanical principles in landslide initiation, both strong moisture penetration by heavy rainfall or rapid snowmelt and the cumulative soil water balance over months or even years are main controlling factors of landslides in low mountain areas of Central Europe (e.g., Krauter and Steingötter 1983; Theisen 1998; Hardenbicker and Grunert 2001; Szabó 2003; Damm 2002; Schmidt and Dikau 2004, Bíl and Müller 2008).

Research on the hydrological causes of landslides today is widely focused on the development of rainfall-related triggering thresholds. The objective thereby is to identify critical levels of rainfall intensity or duration whose exceedance will most likely result in slope failure (cf. Caine 1980; Guzzetti et al. 2008). Trend

M. Klose, *Landslide Databases as Tools for Integrated Assessment of Landslide Risk*, Springer Theses, DOI 10.1007/978-3-319-20403-1_4

and correlation analyses between landslide occurrence and related rainfall or soil moisture conditions are part of a large number of recent studies. Much of these investigations were conducted for study areas located in humid mid-latitude regions (e.g., Chleborad 2003; Jakob and Weatherly 2003; Ibsen and Casagli 2004; Zêzere et al. 2005; Cardinali et al. 2006; Jakob et al. 2006; Guzzetti et al. 2007; Saito et al. 2010). The studies of Crozier (1999), Glade (2000), Godt et al. (2006), Ponziani et al. (2012), and Yeh and Lee (2013) are among those that also consider antecedent moisture conditions and critical soil water contents in the development of rainfall thresholds. Periods of investigation range from a single landslide episode in most cases to a period of 25 years in case of the study of Godt et al. (2006). With regard to Central Europe or related areas worldwide, not only recent landslide monitoring or modeling studies (e.g., Reid et al. 2008; Bell et al. 2010; Jäger et al. 2013; Neuhäuser et al. 2013), but also research works on major rainfall-related triggering events, for instance, that of May 2013 in Germany (cf. Terhorst et al. 2013), have a similar focus on analyzing the correlation between landslide activity and precipitation.

In terms of rainfall thresholds, however, it generally raises the question if soil moisture, which is the true factor of interest in explaining landslide occurrence, can be estimated properly by rainfall data alone (cf. Brocca et al. 2008). Despite this serious concern, there are only few studies that take account of soil moisture estimates (e.g., Ray and Jacobs 2007; Ray et al. 2010; Hawke and McConchie 2011; Brocca et al. 2012), and the modeling of historical soil water balances for their long-term correlation with landslide frequency is still widely exceptional. Lead or initiation times for landslides, expressing the duration of soil saturation to a critical water level, as well as recurrence frequencies of critical soil moisture conditions are rarely available by now, although vital for assessing landslide risk. According to Jordan (1993), Link (1998), Pasuto and Silvano (1998), Terlien (1998), Theisen (1998), and Cardinali et al. (2006) potential initiation times vary between several days or months and periods of one or even 2 years. In contrast with well-established recurrence intervals for triggering rainstorms or rainfall amounts (e.g., Glade 1998; Reid and Page 2002; Salciarini et al. 2008; Frattini et al. 2009), information on the return periods of critical soil moisture levels is except for some rainfall–soil moisture thresholds (cf. Crozier 1999; Glade 2000; Godt et al. 2006; Ponziani et al. 2012) and in situ or remote sensing measurements (e.g., Ray and Jacobs 2007; Ray et al. 2010; Hawke and McConchie 2011) more or less lacking.

The most effective tools to determine hydrologic conditions in soils over prolonged time periods are soil water balance models (e.g., Dyck 1983; Lascano 1991; Xu and Singh 1998). The purpose of applying these models in landslide research is to make use of their capacity to simulate soil water content and level of water saturation as major driving forces of landslide initiation (e.g., Crozier 1999; Yeh and Lee 2013). Generally, three main types of soil water balance models can be differentiated (de Jong and Bootsma 1996): (i) budget models, (ii) semi-dynamic models, and (iii) dynamic models. In studies to assess the influence of soil moisture on landslide initiation, these different model types are mostly applied as lumped-parameter models, considering the root zone of soil as a lumped hydrological system (cf. Ponziani et al. 2012). This means that water transfer between soil, vegetation, and atmosphere is computed using simplified empirical or physical-based relations and model input parameters with a single and regionalized value (cf. Adrien 2004). As a result, lumped-parameter models ignore spatial variation of model input parameters, which contrasts with distributed models embedded in GIS (e.g., Pimenta 2000; Bormann et al. 2009).

The least complex soil water balance models are simple budget models. They are based on the concept of a leaky bucket that fills up to field capacity through precipitation and empties by evapotranspiration and percolation. The soil profile in these models is assumed to be single-layered in most cases. A major advantage of budget models is their minimal data requirement because of evapotranspiration being usually defined as the only unknown in the water balance (e.g., de Jong and Bootsma 1996; Guswa et al. 2002; Zhang et al. 2002). Alternatively, semi-dynamic or dynamic models additionally consider variable infiltration rates and water redistribution in multi-layered soil profiles. Based on further input data on soil hydraulic properties, such models describe the soil hydrological system in greater detail, taking account of water infiltration and the movement of water in unsaturated soil (see also Ranatunga et al. 2008). Besides these different types of soil water balance models, a budgeting of water fluxes is also part of deep percolation or groundwater recharge models (cf. Scalon et al. 2002; Healy 2010), which in their basic form are similar to simple budget models (e.g., Sophocleous 1991).

Selection of an appropriate model for simulating soil moisture requires balancing model complexity against purpose of application and data requirements. It holds true that the higher the level of model complexity, the greater the demand for profound input data (cf. de Jong and Bootsma 1996). A useful criterion when selecting a model is to reduce model complexity to a point at which the model is just able to represent soil hydrological processes with the minimum level of detail required in the application (cf. Zhang et al. 2002). In case of conducting a regional study, Zhang et al. (2002) recommend a "knowledge-based approach" for model parameterization, which is an approach that uses estimation methods to derive water balance components with few, often routinely collected soil, vegetation, and climate data. Much of the previous landslide studies with focus on correlating landslides with soil moisture levels showed such an approach and relied primarily on simple budget or semi-dynamic models (cf. Crozier 1999; Glade 2000; Godt et al. 2006; Ponziani et al. 2012).

4.1.2 Model Description

4.1.2.1 General Overview

In this research, a soil water balance model first conceptualized by Damm (2005) has been optimized to the COupled SIngle-Layer Soil Water Balance MOdel

(COSIMO). COSIMO is a simple budget model in the form of a lumped-parameter model that is developed as modeling tool in MS Excel. This type of model was chosen as its purpose of application is to gather regionalized overview data on long-term soil moisture trends. There is thus the need for strong simplification of soil hydrological processes, which can be best achieved by the use of a simple budget model. The main idea of model development was to combine and modify existing calculation methods in a way that enables estimation of regional soil moisture conditions with low data requirements. The advantage of COSIMO is thus to run with few input data that are readily available at regional level. COSIMO includes calculation methods tailored for use in low mountain areas of Central Europe or regions with comparable landscape and climate conditions (cf. Klose et al. 2012a). The model parameters have been specified for characteristic types of soils and vegetation at landslide sites in the Upper Weser area (cf. Sect. 3.2).

The model architecture of COSIMO consists of two main components that enable simulation of soil moisture in two different steps (Fig. 4.1). In the first step, potential evapotranspiration is estimated using a combination of three coordinated

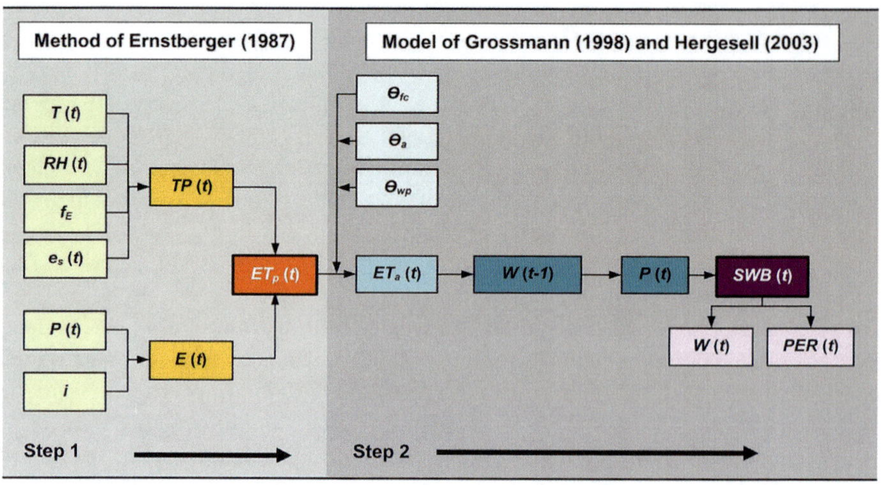

$T(t)$:	Air temperature [C°]	$ET_p(t)$:	Potential evapotranspiration [mm]
$RH(t)$:	Relative humidity of the air [%]	$ET_a(t)$:	Actual evapotranspiration [mm]
f_E:	Plant coefficient, Ernstberger (1987)	Θ_{fc}:	Soil water content at field capacity [mm]
$e_s(t)$:	Saturated vapor pressure of the air [hPa]	Θ_a:	Available water capacity [mm]
$P(t)$:	Precipitation [mm]	Θ_{wp}:	Soil water content at wilting point [mm]
i:	Interception factor, Brechtel and Pavlov (1977)	$SWB(t)$:	Soil water balance [mm]
$TP(t)$:	Potential transpiration [mm]	$W(t)$:	Soil water content [mm]
$E(t)$:	Potential canopy evaporation [mm]	$PER(t)$:	Percolation [mm]
		t:	Time period [month or day]

Fig. 4.1 Model architecture of the soil water balance model COSIMO (COupled SIngle-Layer Soil Water Balance MOdel) with the main model parameters and calculation steps. The modeling approach combines the method of Ernstberger (1987) for calculation of potential evapotranspiration with the regional groundwater recharge model from Grossmann (1998) and Hergesell (2003). COSIMO is suitable for regional simulation of soil moisture in humid low mountain areas of the mid-latitudes such as large parts of Central Europe (*Source* modified after Klose et al. 2012a)

calculation methods, including a modified version of the method of Ernstberger (1987). This method for calculation of potential evapotranspiration constitutes an upgrade of the equation of Haude (1955), whereby the modifications made mainly refer to the parameterization of canopy evaporation by a simple empirical relation according to Brechtel and Pavlov (1977). The second step of this modeling approach involves simulation of soil moisture based on the estimates of potential evapotranspiration that are derived in the first step (cf. Klose et al. 2012a). Both the soil water balance equation and the algorithm for simulation of soil moisture are based on a method for estimating regional groundwater recharge that was first developed by Grossmann (1998) and Hergesell (2003). In the following, the modeling approach is presented in detail, including calculation steps, parameter specification, and model validation.

4.1.2.2 Calculation Steps

The following soil water balance equation is at the heart of COSIMO and defines the model parameters that need to be specified (modified after Grossmann 1998 and Hergesell 2003):

$$SWB(t) = \mathrm{W}(t-1) + P(t) - ET_a(t) \tag{4.1}$$

where $SWB(t)$ is the soil water balance, $W(t)$ is the soil water content, $P(t)$ is the precipitation, and $ET_a(t)$ is the actual evapotranspiration. The time period t is variable and can be days or months, with no substantial difference between operating on a daily or monthly basis. The advantage of using $ET_a(t)$ instead of the potential evapotranspiration $ET_p(t)$ is to consider reduced transpiration during dry periods. $ET_a(t)$ is derived from $ET_p(t)$ after calculation of $ET_p(t)$ on the basis of the method of Ernstberger (1987), which in modified form is as follows:

$$ET_p(t) = E(t) + TP(t) \tag{4.2}$$

where $E(t)$ is the potential canopy evaporation and $TP(t)$ is the potential transpiration, both of which given for a specific vegetation type. The calculation of $TP(t)$ uses the following relation that is derived from the methods of Haude (1955) and Ernstberger (1987):

$$TP(t) = f_E \cdot \left(e_s(t) \cdot \left(1 - \frac{RH(t)}{100}\right)\right) \cdot d(t) \tag{4.3}$$

in which f_E is a plant coefficient after Ernstberger, $e_s(t)$ is the saturated vapor pressure of the air, and $RH(t)$ is the relative humidity of the air, and $d(t)$ is the number of days of month t. The parameter f_E is an empirical field constant that replaces the Haude factor in this approach. According to the Magnus formula (cf. Kraus 2004), $e_s(t)$ is given by

$$e_s(t) = 6.107 \cdot 10^{\frac{a \cdot T(t)}{b+T(t)}} \tag{4.4}$$

with the constants

$$\begin{cases} a = 7.5 \text{ and } b = 235 & T(t) > 0 \\ a = 7.6 \text{ and } b = 240.7 & T(t) \leq 0 \end{cases} \tag{4.5}$$

where *T*(*t*) is the air temperature. Alternatively, *E*(*t*) is computed on the basis of the following empirical relation that is modified after Hergesell (2003):

$$E(t) = P(t) \cdot \frac{i}{100} \tag{4.6}$$

in which *i* is an interception factor according to Brechtel and Pavlov (1977). The interception factor describes the percentage of intercepted precipitation for various tree species for both growing season (April to October) and non-growing season (November to March). As calculated in this approach, *E*(*t*) represents the total unproductive evaporation, including canopy and litter interception loss. The values calculated for *TP*(*t*) and *E*(*t*) enter Eq. (4.2) to obtain $ET_p(t)$, and after having determined this variable, the first step of COSIMO is completed (cf. Klose et al. 2012a). In comparison to related calculation methods (e.g., Penman 1948; Blaney and Criddle 1962; Monteith 1976), the approach of Ernstberger (1987) has the advantage of being developed for German low mountain areas and of requiring relatively few meteorological input data.

Step two of COSIMO refers to the simulation of soil moisture using a simplified version of the groundwater recharge model of Grossmann (1998) and Hergesell (2003). Based on their modeling approach, $ET_a(t)$ is calculated as follows:

$$ET_a(t) = \begin{cases} ET_p(t) & W(t-1) \geq \theta_a \cdot 0.6 + \theta_{wp} \\ (W(t-1) - \theta_{wp}) \cdot \frac{0.6 \cdot ET_p(t)}{\theta_a} & W(t-1) < \theta_a \cdot 0.6 + \theta_{wp} \end{cases} \tag{4.7}$$

where θ_a is the available water capacity and θ_{wp} is the soil water content at wilting point. The algorithm of soil moisture simulation derived from this groundwater recharge model is given by

$$SWB(t) = W(t-1) + P(t) - ET_a(t) \tag{4.8}$$

and

$$\theta(t) = SWB(t) = W(t) + PER(t) \tag{4.9}$$

with

$$PER(t) = \begin{cases} 0 & SWB(t) \leq \theta_{fc} \\ SWB(t) - \theta_{fc} & SWB(t) > \theta_{fc} \end{cases} \tag{4.10}$$

and

$$SWB(t = 0) = \theta_{fc} \tag{4.11}$$

in case of starting the simulation in winter, where *PER*(t) is the percolation, θ_{fc} is the soil water content at field capacity, and $\theta(t)$ is the soil moisture. Putting the initial value of *SWB*(t) equal to θ_{fc} means that the soil is assumed to be completely saturated, which is a useful assumption when starting the model simulation in winter months (November to March). In subsequent time steps, *SWB*(t) is calculated as function of the soil water content of the previous time step (month or day), and percolation or groundwater recharge only takes place if the soil is completely saturated, implicating exceedance of field capacity. The soil moisture can be best represented in terms of *SWB*(t) that includes both pore water and percolation water. As compared with *W*(t), a key advantage of *SWB*(t) is its capability to describe oversaturation in the soil, which is critical for model application in landslide studies (cf. Klose et al. 2012a).

4.1.2.3 Specification of Model Parameters

The application of COSIMO makes it necessary to specify the model parameters of above calculation methods. A main idea of COSIMO is to perform parameter specification to a large extent by desk studies, including, amongst others, analysis of routine climate data or data tables of plant coefficients (cf. Klose et al. 2012a). Soil and vegetation data in this research have been gathered from the available landslide database. The present study applies COSIMO on a monthly basis for the period 1953–2011. This period was chosen because of the availability of a widely complete landslide record at local level since about the early 1950s. Temporal model parameters of COSIMO were thus specified by using input data derived from monthly data products. The input data for customization of COSIMO originate either from the Upper Weser area or regions in adjacent areas of the German Federal State of Hesse that largely allow comparison and reasonable transfer of data (cf. Klose et al. 2012a).

Climate data incorporated in the model refer to meteorological station Kassel (city of Kassel, northern Hesse) from the German Meteorological Service (DWD). As is the case with the next nearest station Göttingen, which shows a similarly long record, the station Kassel is located in a basin area, wherefore the climate conditions are expected to be somewhat drier and warmer than in the Upper Weser area, which is situated 15 km northeast. However, as monthly mean values are considered in this analysis, the climate record is smoothed compared to daily rainfall extremes, thus approximating the general rainfall pattern, even though being not fully representative (cf. Klose et al. 2012a). The station Kassel shows a widely complete climate record with only few data gaps that were filled by using monthly average values. Temperature data and the data for relative humidity refer to the measurement time at 2:30 pm CET (12 am UTC since 09-2004). The reason for choosing this measurement time instead of the daily mean was to avoid underestimation of transpiration caused by reduced temperature or humidity in the evening or early morning hours (see also Damm 2005). All climate data used in this study can be downloaded as monthly data products from the DWD website (http://www.dwd.de/).

The available landslide database stores information on types of vegetation at landslide sites in the Upper Weser area. Database analysis revealed that Fagus sylvatica usually is the dominant tree species at most landslide sites where it often occurs in pure or mixed stands with intermediate to high stand ages (cf. Klose et al. 2012a). The plant coefficient f_E in Eq. (4.3) therefore corresponds to the values given by Ernstberger (1987) for old beech stands (>80 years). This also applies to the interception factor i that was transferred from Brechtel and Pavlov (1977) to determine potential canopy evaporation in Eq. (4.6). Besides specification of vegetation parameters, the application of COSIMO requires defining the values of different soil hydrological parameters, including soil water content at field capacity (θ_{fc}) or wilting point (θ_{wp}) and available water capacity (θ_a). Specification of soil-related input data is based on characteristic values of regional soil types that are typically affected by landslides. Most commonly, according to landslide database information, these are slightly sandy loam, loamy sand, and silty loamy sand, with each of that soil types showing moderate bulk density. The depth of the root zone corresponds to that of typical forest soils and thus was estimated at 120 cm on average (cf. Klose et al. 2012a). This database information along with data tables of AG BODEN (2005) support specification of required soil hydrological parameters for being able to run the soil moisture simulation with COSIMO.

4.1.2.4 Model Calibration and Validation

The modeling approach from Grossmann (1998) and Hergesell (2003) was primarily developed and calibrated on the basis of discharge and lysimeter measurements. Besides comparison of the modeling results with field data, the model validation in Hergesell (2003), for instance, involved cross-checking of simulation data with exemplary results of further groundwater recharge studies (e.g., HLfU 1995) and published infiltration or percolation rates (e.g., Arbeitskreis Grundwasserneubildung FH-DGG 1977; Hölting 1996). Given these first plausibility tests, the applied model is proven to be well-suited for regional groundwater recharge studies in Central Europe (cf. Hergesell 2003). In the present investigation, by contrast, direct validation of the obtained simulation results was not possible. The main reasons were lack of in situ soil moisture measurements and general absence of suitable reference data (cf. Klose et al. 2012a). However, it is referred at this point to the validation of Hergesell (2003) who compared his groundwater recharge estimates for Hesse with annual percolation rates derived from lysimeter stations distributed throughout this state. Even though there are considerable conceptual limitations in this kind of validation, a mean deviation of −6 % between model and field data indicates satisfying model accuracy at least to some extent. Nevertheless, the accuracy assessment performed by Hergesell (2003) must be considered critically, and the use of these validation results to evaluate the plausibility of the model presented in this work is insufficient for a robust and representative model validation (cf. Klose et al. 2012a). Further research is therefore needed to be able to perform a consistent validation of COSIMO and to check the accuracy of this model in landslide-related soil moisture simulations. For more detailed information on the difficulties and limitations of the model validation, it is also referred to the cited works.

4.2 Landslide Susceptibility Model

4.2.1 Introduction

Hillslopes around the world are susceptible to landslides and thus their development creates serious risk to people and property (e.g., Brabb 1991; Dilley et al. 2005; Nadim et al. 2006; Hong et al. 2007; Kirschbaum et al. 2010; Petley 2012). In order to promote hazard awareness in hillslope development, landslide susceptibility models, which enable identification, mapping, and zonation of landslide-prone areas (e.g., Chacón et al. 2006; Corominas et al. 2014), play an increasingly important role in management and reduction of landslide risk worldwide (e.g., Jones 1992; Cascini et al. 2005; Lateltin et al. 2005; Schwab et al. 2005; Schuster and Highland 2007; Fell et al. 2008; Wang et al. 2013). As defined by Van Westen et al. (2006), landslide susceptibility describes the spatial probability of landslide occurrence, and Aleotti and Chowdhury (1999) argue that susceptibility assessment requires clarifying "*where, what types of, and how landslides will occur*". The main idea of landslide susceptibility modeling, according to Brenning (2005), is to predict the potential location of future landslides by analyzing presence or absence of geofactors existing at sites where landslides occurred in the past.

A large number of natural and human factors of dynamic or quasi-static nature have an effect on landslide susceptibility (e.g., Alexander 1992; Popescu 1994; Hutchinson 1995; Soeters and Van Westen 1996; Crosta et al. 2012). These factors are also known as predisposing (quasi-static; e.g., soil properties) and/or preparatory factors (dynamic; e.g., soil moisture), both below referred to as predisposing factors. In contrast to factors that trigger landslides, predisposing factors destabilize slopes to a marginally stable level, thus increasing probability of slope failure but without causing landslide activity (cf. Crozier 1986; Glade and Crozier 2005; GuiYun et al. 2008; Van Westen et al. 2008). Besides slope gradient and slope aspect as major geomorphic predisposing factors, lithology, soil material, and vegetation are important static predisposing factors that are frequently considered in landslide susceptibility modeling (e.g., Corominas et al. 2014). Most investigations today (see below) concentrate on such geospatial data of static nature which are comparatively easy to obtain and still show high explanatory power (cf. Fressard et al. 2014; Klose et al. 2014a). The many predisposing factors and their complex effects on slope stability, however, generally require for strongly simplified modeling approaches (cf. Carrara et al. 1999), and important dynamic influences, for instance, that caused by temporally variable land use practices, are inevitably ignored in most of today's landslide susceptibility models (cf. Van Beek and Van Asch 2004; Van Westen et al. 2008). It is therefore necessary to bear in mind that in modeling of landslide susceptibility there is great uncertainty always existent (Guzzetti et al. 2006; see also Klose et al. 2014a).

Many different approaches and methods are currently available to model landslide susceptibility by means of GIS software. All of these approaches and methods have in common the following basic assumptions proposed by Varnes and IAEG (1984):

(i) "The past and present are keys to the future". This assumption implies that landslides in future will most likely occur in identical physiographic settings than in the past and present. Knowledge on the location of past and present landslides is thus vital for prediction and zonation of areas that are susceptible to landslides in future.

(ii) "The main conditions that cause landsliding can be identified". Hutchinson (1995) upgrades this assumption and notes that "*the main conditions that cause landsliding are controlled by physical factors and are therefore, in principle, identifiable*". This assumption enables to develop models of landslide susceptibility by referring to predisposition factors which can be identified based on their spatial association with previous landslides. These predisposition factors are assumed to physically control landslide susceptibility and enable hazard identification over broader areas according to their spatial presence or absence (see above).

(iii) "Degrees of hazard can be estimated". As specified by this assumption, the relative contribution of each predisposing factor to past and present landslide occurrence can be expressed in a qualitative or quantitative measure that can be determined empirically, statistically or deterministically (see also Guzzetti et al. 1999). Landslide susceptibility in an area with a certain spatial configuration of predisposing factors can thus be estimated by combining the measures derived for relevant predisposing factors in the area under consideration.

Additionally, Hutchinson (1995) states a further assumption, which is as follows:

(iv) "The various types of landsliding can generally be recognized and classified, both morphologically, geologically and geotechnically". This assumption is of major importance for the development of spatial landslide inventories as basis for modeling of landslide susceptibility. A comprehensive overview of most recent techniques of landslide recognition and mapping for the purpose of developing landslide inventories is given by Guzzetti et al. (2012).

The approaches available for landslide susceptibility modeling can be classified in three main categories, including (i) heuristic approaches, (ii) inventory approaches, (iii) statistical approaches, and (iv) deterministic approaches (Table 4.1). Heuristic approaches are also known as direct approaches in which landslide susceptibility modeling is based on expert opinion. All other approaches constitute indirect approaches. They predict landslide susceptibility by means of statistical or deterministic models, wherefore often considered to be more objective than direct approaches that rely on personal experience. Thus, direct approaches show a qualitative character, while indirect ones are referred to as quantitative approaches. The advantage of quantitative approaches is seen in their capacity to provide reproducible numerical estimates of landslide susceptibility. For the above and further information on available modeling approaches and methods, it is referred to the technical reviews from Soeters and Van Westen (1996), Guzzetti et al. (1999), Van Westen (2000), Aleotti and Chowdhury (1999), Parise (2001), Dai et al. (2002), Wang et al. (2005), Chacón et al. (2006), Hervás and Bobrowsky (2009), Kanungo et al. (2009), Pardeshi et al. (2013), and Corominas et al. (2014).

The first step in landslide susceptibility modeling usually is to select an appropriate modeling approach. According to Aleotti and Chowdhury (1999), model selection depends on the purpose of application, the size of the study area, and data availability. Inventory and statistical approaches are suited for use at regional to national level, whereas application of deterministic approaches is restricted to studies with focus on modeling the stability of single slopes (Table 4.1). Most common approaches of inventory analysis are landslide location, isopleth or density maps that are able to inform about the spatial distribution and frequency of past landslides over broad areas (e.g., Wright et al. 1974; DeGraff and Canuti 1988; Bulut et al. 2000; Galli et al. 2008). Such maps are the simplest types of hazard assessments, with density maps, for instance, assuming landslides to be a spatially continuous variable, thus showing the best results in case of homogenous

Table 4.1 General summary of existing approaches and methods for mapping of landslide susceptibility with an overview of recommended scales of use (*Source* modified after Soeters and Van Westen 1996 and updated according to Gokceoglu and Sezer 2012 and Corominas et al. 2014)

Approaches	Methods	Scale of use		
		Regional (1:100,000)	Medium (1:25,000)	Large (1:10,000)
Inventory	Landslide inventory maps	Yes	Yes	Yes
	Landslide density maps	Yes	No	No
Heuristic	Geomorphological mapping	Yes	Yes	Yes
	Qualitative map combination	Yes	Yes	No
Statistical	Bivariate	Yes	Yes	No
	Information Value method			
	Weights-of-Evidence model			
	Multivariate	Yes	Yes	No
	Discriminant analysis			
	Logistic regression			
	Soft computing	Mostly medium scales		
	Fuzzy logic approach			
	Artificial Neural Networks			
Deterministic	Slope stability models	No	No	Yes

landscapes (Guzzetti et al. 2000). Given these deficits in mapping presence and spatial variation of predisposing geofactors, inventory approaches, if not considering geomorphological slope units (cf. Carrara et al. 1995), tend to provide ambiguous results (see also Guzzetti 2005), wherefore a statistical approach is chosen in this study (cf. Klose et al. 2014a).

Most landslide susceptibility models developed today are based on statistical methods, including bivariate or multivariate approaches and new modeling concepts from soft computing (Table 4.1). Statistical methods are principally applicable on regional scale (Corominas et al. 2014), but when considering purpose of application and data availability, not all of them are equally suitable. Thus, new methods from soft computing such as fuzzy logic or artificial neural networks, enabling to handle nonlinear landslide patterns or complex factor interaction (e.g., Kanungo et al. 2006, 2009), require a high degree of specialization, and complex data requirements limit their application, especially at larger scales (cf. Gokceoglu and Sezer 2012). The most frequently used methods of the statistical approach are, according to Brenning (2005), multivariate methods, in particular logistic regression and discriminant analysis. Multivariate methods consider correlation of predisposing factors by performing multivariate statistics on the basis of a data table of mapping unit values (e.g., Lee and Min 2001; Ohlmacher and Davis 2003; Santacana et al. 2003; Ayalew and Yamagishi 2005; Van Den Eeckhaut et al. 2006; Mathew et al. 2009; Mancini et al. 2010). The application of multivariate methods, however, is subject to similar restrictions as soft computing, which is why these methods are ignored in method selection as well (cf. Klose et al. 2014a).

The idea of bivariate methods is to use measures of landslide density derived from map combination to calculate a numerical weight for each predisposing factor considered in the model (e.g., Van Westen 1993; Süzen and Doyuran 2004a; Magliulo et al. 2008). Two different methods are most commonly applied to determine factor weights, which are the Information Value approach (Kobashi and Suzuki 1988; Yin and Yan 1988) and the Weights-of-Evidence model (Bonham-Carter et al. 1989). Compared to the Weight-of-Evidence model (see also Neuhäuser et al. 2012a, b), factor weighting based on the Information Value approach is somewhat simpler, with each factor weight being simply expressed as log of a density ratio (cf. Sect. 4.2.2). In contrast to frequent comparison of bivariate and multivariate methods (e.g., Süzen and Doyuran 2004b, Wang and Sassa 2005; Nandi and Shakoor 2009; Shahabi et al. 2013), the performance of the two bivariate methods has not been compared so far, wherefore little is known about which bivariate method is better adapted for use at regional level. Given the data situation for Lower Saxony, the flexibility of the Information Value method with regard to customization, however, is seen as key advantage of this modeling approach, which justifies its applications in the present study (cf. Klose et al. 2014a). The Information Value method has already been applied in a large number of investigations at both local and regional scale for different areas worldwide (e.g., Wu et al. 2001; Zêzere 2002; Çevik and Topal 2003; He and Beighley 2008; Magliulo et al. 2008; Yalcin 2008; Conforti et al. 2012; Guillard and Zezere 2012).

4.2.2 *Model Description*

4.2.2.1 General Overview

In this study, a landslide susceptibility model for Lower Saxony has been developed by using a modified Information Value approach (cf. Klose et al. 2014a). This bivariate statistical method calculates a factor weight denoted as information value for each predisposing factor considered in the model. There are principally no strict rules but guidelines and test procedures available to perform the selection of relevant predisposing factors (e.g., Van Westen et al. 2008; Costanzo et al. 2012; Pereira et al. 2012; Corominas et al. 2014). According to Van Westen et al. (2008), factor selection depends, amongst others, on scale of analysis, characteristics of study area, and landslide types. A key role is thereby played by scale of analysis because important input data, for example, data on lithology or land use, are often restricted to data products with small spatial resolution. This data limitation conflicts with model applications that are focused on analyzing landslide susceptibility at regional to national scale (cf. Klose et al. 2014a). As generally is the case with landslide hazard studies, input data of modeling approaches are like their formulization always less than perfect, and time and/or cost restrictions usually require simplifications (cf. Turner and McGuffey 1996).

Landslide susceptibility modeling using GIS and bivariate statistics is based on a specific procedure that follows a predefined workflow with a series of tasks related to collection, preparation, and analysis of geospatial data and landslide information (Fig. 4.2). There are always several milestones to pass in a systematic assessment of landslide susceptibility, including data compilation, generation of input data, susceptibility modeling, map creation, and model validation (see also Van Westen 1993; Aleotti and Chowdhury 1999). These milestones also define the workflow of model development in the present study (cf. Klose et al. 2014a). All working steps in landslide susceptibility modeling today are supported by the application and functionality of GIS software and the broad spectrum of GIS-based data products (e.g., Van Westen 2000, 2004). The software used in this investigation is ESRI ArcView GIS 10.0.

4.2.2.2 Input Data

Landslide Inventory

Basic requirement for landslide susceptibility modeling is the availability of a spatial landslide inventory. The modeling approach developed in Klose et al. (2014a) uses an inventory that has been created from a regional subset of the available landslide database and from data sets of the landslide distribution map published by Schunke (1971). This landslide distribution map displays in paper format the location and spatial extent of landslide areas at cuesta scarps in the Weser-Leine

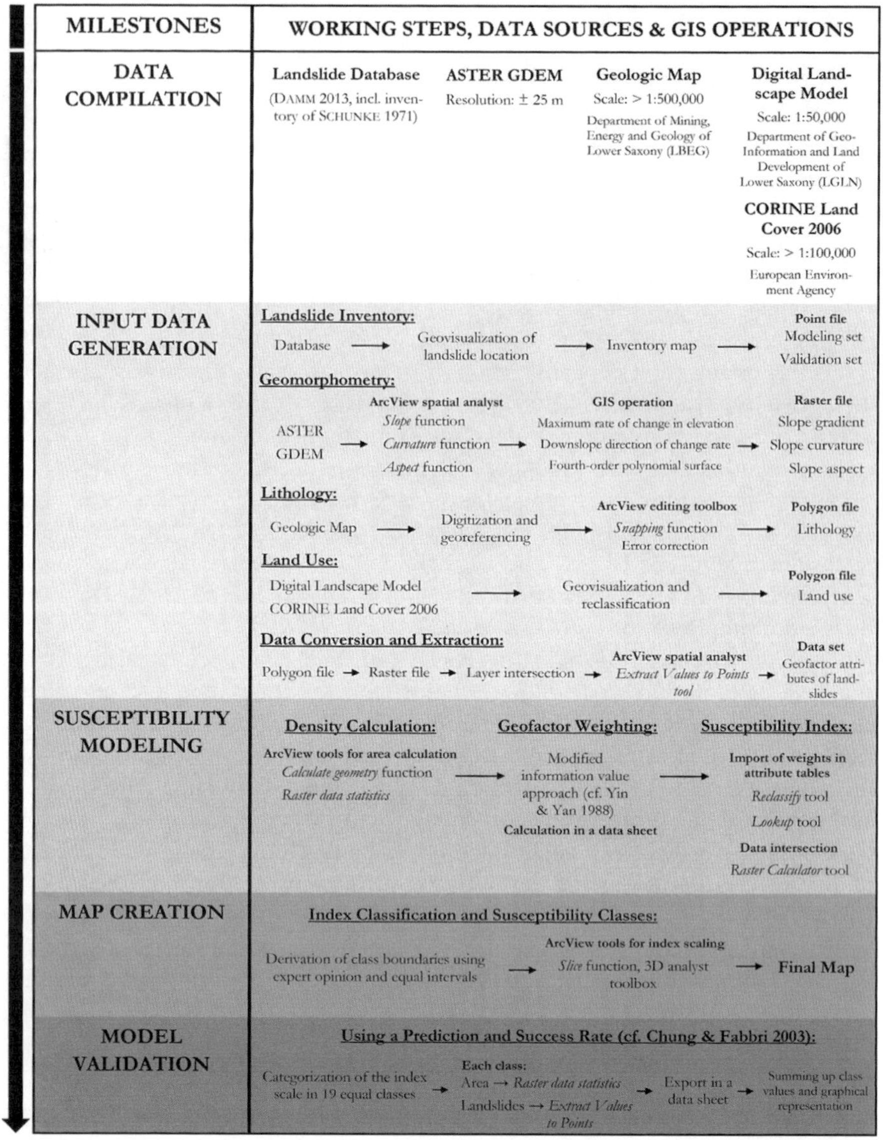

Fig. 4.2 Overview of the milestones and working steps in the development of the landslide susceptibility model for Lower Saxony. The flowchart also illustrates the input data for model development and some important GIS operations using the software ESRI ArcView GIS 10.0 (*Source* Klose et al. 2014a)

Uplands at 1:200,000 scale. It further differentiates landslide occurrence date (i.e., pleistocene, holocene or recent landslide) and illustrates the types of lithology forming the main scarps in this region. Major disadvantage of the landslide distribution map is its rough manual delineation of landslide areas which implies serious problems for data digitization. In order to avoid errors and inaccuracy caused by the transfer of landslide data, geocoding and mapping of the exact landslide location was assisted by expert knowledge, literature information, and the use of Google Earth imagery. Geographic coordinates of landslide areas extracted from this landslide distribution map were reduced to a single point location whose spatial reference was determined on the basis of the position of the main scarp or the top of the displaced mass. The data sets on geographic landslide location gathered either from the database or the distribution map were stored in a separate database maintained as MS Excel spreadsheet. On the basis of the location data stored in this database, a spatial landslide inventory has been compiled by importing these landslide data in ArcView GIS. The developed spatial landslide inventory includes location data (geographic coordinates, landslide site, administrative region) for 889 landslide sites in Lower Saxony, especially the Lower Saxon Uplands (Fig. 4.3; cf. Klose et al. 2014a).

Geomorphometry

This modeling approach uses digital data sets on slope gradient, slope curvature, and slope aspect to take account of topographic relief as major predisposing factor of landslides (cf. Klose et al. 2014a). The considered geomorphic terrain data have been extracted and calculated from the ASTER Global Digital Elevation Model (ASTER GDEM) that is available online at the LP DAAC data center of NASA and USGS (http://gdex.cr.usgs.gov/gdex/). This open-source DEM has a spatial resolution of ±25 m and is due to its comparatively high spatial accuracy of general acceptance in regional landslide susceptibility modeling (cf. Oh et al. 2012; Van Westen et al. 2008). Much of the relief analysis was performed using different ArcView spatial analyst tools that enable calculation of relevant geomorphic terrain parameters (Fig. 4.2). The calculated data set on slope gradient was categorized in seven attribute classes with intervals of 7° and one attribute class for slope gradients between 49° and 90°. Slope gradients >49° have been merged in one class due to the fact that no landslide is recorded above this value. Alternatively, classification of slope aspect follows the main directories, including a further category for flat areas. Slope curvature is considered in the form of plan curvature, with values ranging between –420 and 720, where negative or positive values indicate concavity or convexity. A majority of the curvature data concentrates between the values –2 and 2, wherefore these values represent the upper boundaries for the classes concave and convex. All other values were categorized to classes symbolizing strong concave or strong convex curvature or to the class for slopes that show no curvature (cf. Klose et al. 2014a).

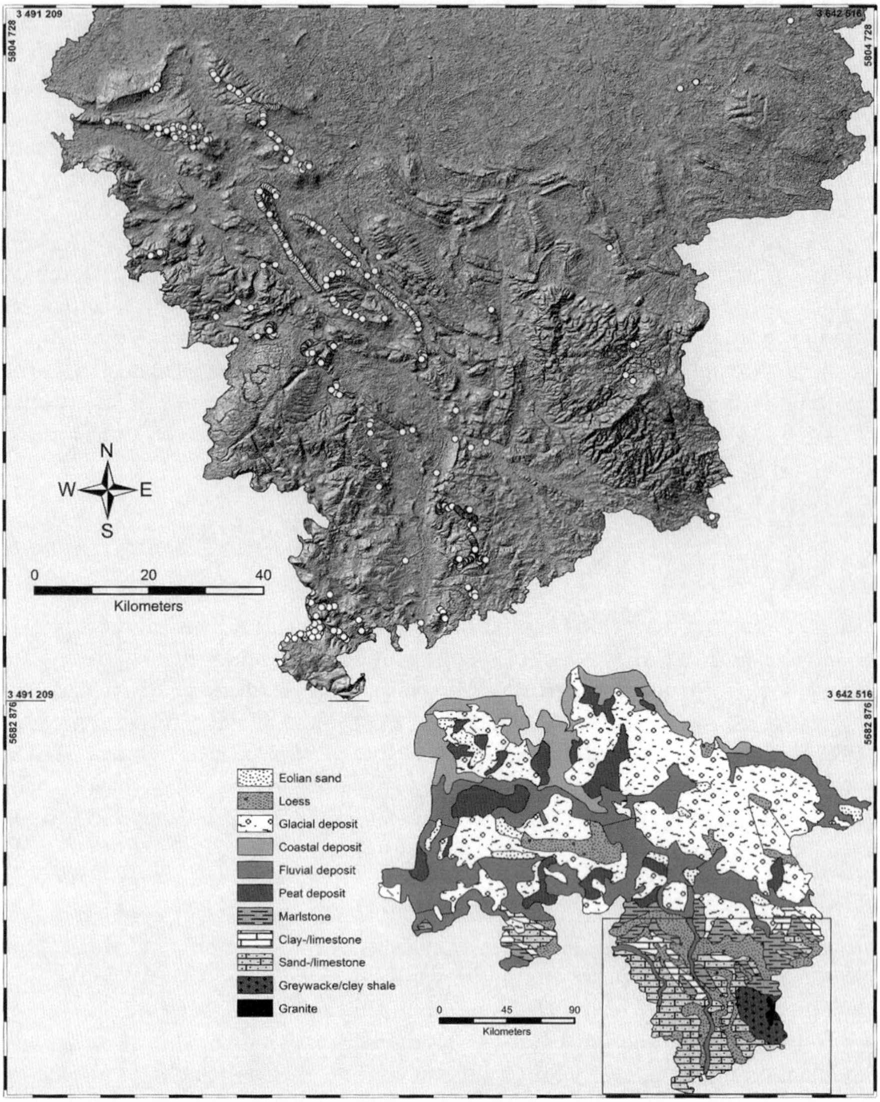

Fig. 4.3 Spatial landslide inventory for the Lower Saxon Uplands (*upper part*) in southern Lower Saxony. The inventory map reveals a spatial clustering of landslides along the main scarps and ridges in the Weser-Leine Uplands. A further cluster of landslide activity is represented by the deep cut river valleys of the Upper Weser area. Only few landslides have been identified and mapped for the Harz Mountains which were not in the focus of previous data collection. The figure also shows the generalized geologic map (*lower part*) that provided the lithological input data for the developed landslide susceptibility model. This map was digitized from a derivate of a geologic overview map at 1:500,000 scale (*Source* Klose et al. 2014a)

Lithology

Information on lithology is gathered in this study from a generalized geologic map provided by the Lower Saxony Department of Mining, Energy and Geology (LBEG). The map is available online (http://www.lbeg.niedersachsen.de/) and constitutes a derivate of a geologic overview map at 1:500,000 scale. It differentiates besides two lithological classes for Paleozoic basement (granite, greywacke/ clay shale), three main classes of Mesozoic and Tertiary sedimentary rock, including sand-/limestone, clay-/limestone, and marlstone. Furthermore, the map illustrates several types of Quaternary lowland deposits, which were summarized in a separate attribute class. As this map is not available in GIS format, its digitization was necessary to obtain a vector data set of lithology supporting GIS-based data processing. Errors in data digitization, for instance, data gaps or overlapping features, were corrected using tools from the ArcView GIS editing toolbox (Fig. 4.2). The result of data digitization and editing is a polygon shapefile storing spatial data on the main types of lithology in Lower Saxony (cf. Klose et al. 2014a).

Land Use

Two different information sources of varying thematic complexity were analyzed to integrate land use data into the landslide susceptibility model. Most data sets on land use have been extracted from the official Lower Saxony Digital Landscape Model (DLM 50) developed and provided by the Lower Saxony Department of Geoinformation (LGLN). This digital data product for basic geospatial information reproduces the land surface and its main objects in high thematic precision and provides differentiated landscape data for topographic maps with a scale of 1:50,000. For representation of land use conditions in adjacent areas of northern Hesse, the data included in the DLM 50 were complemented by land use information derived from CORINE Land Cover 2006 (CLC2006; available at http://www.eea.europa.eu/). Both data sets have a specific thematic layer structure and differ in their representation of topographic objects. In order to combine information from both sources in consistent object classes, data reclassification was necessary. This especially applied to some main types of forest vegetation, which were grouped in one single category. The developed landslide susceptibility model considers in total five different land use classes, including forest, grassland, arable land, urban area, and "other land use type". The latter category includes besides roads and water bodies, areas that are not further specified (cf. Klose et al. 2014a).

4.2.2.3 Modeling Approach

Factor Weighting

Factor weighting and combination of factor weights later on requires data sets with same spatial reference and resolution. The basic mapping unit in this study is the grid cell of the ASTER GDEM. All other data sets were transformed to this raster format except for the landslide inventory map which was used as basic data layer for the intersection of factor maps (lithology, land use, slope gradient, etc.). Identification of relevant factor attribute classes existing at each landslide site was supported by the use of tools for data extraction provided by the ArcView GIS spatial analyst (Fig. 4.2). The extracted data sets were exported to a data matrix managed in MS Excel that stores the combinations of factor attribute classes found at the different landslide sites. This data matrix was used to perform factor weighting by means of the Information Value approach in a separate MS Excel spreadsheet application (cf. Klose et al. 2014a). The modeling technique applied for geofactor weighting calculates an information value I as numerical weight for each geofactor attribute $A(i)$ (Kobashi and Suzuki 1988; Yin and Yan 1988). The value of $I_{A(i)}$ describes the contribution of $A(i)$ to former landslide occurrence and can be expressed in terms of probability as follows (cf. Wu et al. 2001; Wang and Sassa 2005):

$$I_{A(i)} = \ln \frac{P\{B|A(i)\}}{P\{B\}} \qquad (i = 1, 2, 3, \ldots, n) \tag{4.12}$$

where $P\{B \mid A(i)\}$ is the landslide probability in the presence of $A(i)$ and $P\{B\}$ is the overall landslide probability. Since a probability concept is problematic in a data-driven approach, frequency statistics are used to specify this probabilistic relationship. Therefore, Eq. (4.12) is converted to an expression of the following form (modified after Yin and Yan 1988):

$$I_{A(i)} = \ln \frac{N_{A(i)}/S_{A(i)}}{N/S} \qquad (i = 1, 2, 3, \ldots, n) \tag{4.13}$$

where $N_{A(i)}$ is the number of landslides in attribute class $A(i)$, N is the number of landslides in the entire territory, $S_{A(i)}$ is the total area of attribute class $A(i)$, and S is the total area of the entire territory. This means that $I_{A(i)}$ is calculated by dividing the landslide density of a certain factor attribute by that of the entire study area. As result of taking the natural logarithm, $I_{A(i)}$ is positive or negative if the landslide density of $A(i)$ is greater or less than the average landslide density (cf. Van Westen 2000). Factor attributes showing positive values of $I_{A(i)}$ are thus likely to promote instability, while otherwise their influence is assumed to be of stabilizing nature. It holds true that the higher the value of $I_{A(i)}$ the stronger this relationship (cf. Zêzere 2002; Wang and Sassa 2005).

Susceptibility Index

In order to perform spatial modeling of landslide susceptibility, the values of $I_{A(i)}$ calculated in the MS Excel application needed to be reimported to ArcView GIS. This was done by copying each information value to the attribute table of the corresponding factor attribute map. By means of reclassification tools from the ArcView GIS spatial analyst (Fig. 4.2), which enabled to write the information values to separate data raster, weight maps were created for the different attribute classes of each considered geofactor. The next modeling step involved combination of weight maps to determine the susceptibility index $SI(x)$ that describes the landslide predisposition of the basic mapping unit (cf. Klose et al. 2014a). This index is defined as the sum of all values of $I_{A(i)}$ in a defined grid cell x and can be computed as follows (modified after Yin and Yan 1988):

$$SI(x) = \sum_{i=1}^{n} \ln \frac{N_{A(i)}/S_{A(i)}}{N/S} \tag{4.14}$$

The calculation of $SI(x)$ was performed using the *Raster Calculator* that is embedded in ArcView GIS and that supports data intersection of overlaying weight maps. The result of this data intersection, or more specifically, this grid-based adding of information values, was a new raster file referred to as susceptibility index map. This map displays a specific value of $SI(x)$ for each grid cell x of the ASTER GDEM data raster (cf. Klose et al. 2014a).

Hazard Classes

The obtained susceptibility index map illustrates the level of landslide susceptibility on a continuous scale of numerical values. Such a data representation causes difficulties in interpretation and requires simplification for proper hazard communication. It is therefore necessary to categorize the range of values of $SI(x)$ in several hazard classes, whereby a differentiation of three to six classes is most commonly used today (cf. Ayalew et al. 2004; Beguería 2006; Chung and Fabbri 2003). Scaling of hazard classes can be realized by applying various statistical data classifiers such as natural breaks, quantiles, equal intervals, and standard deviation. Despite the availability of statistical classification systems, it is still common practice in most studies to perform scaling of hazard classes based on expert opinion (cf. Ayalew et al. 2004). The same applies to this study in which an expert-based classification was combined with equal intervals as both techniques in combination best enabled to capture the main data patterns within the hazard scale (cf. Klose et al. 2014a). For the definition of reasonable hazard classes, the influence of various factor combinations on the value of $SI(x)$ was tested, with the objective to identify natural break points that serve as empirical class boundaries. In cases where class boundaries could not be set logically, equal intervals were used to group the data in a systematic way. After determining the different class boundaries, the susceptibility index map was reclassified with ArcView GIS tools for data grouping (Fig. 4.2) to obtain the final landslide susceptibility map (Fig. 5.1).

4.2.2.4 Model Validation

The objective of model validation is to test both accuracy and predictive power of the developed landslide susceptibility model (e.g., Remondo et al. 2003; Beguería

2006; Van Den Eeckhaut et al. 2010). Model accuracy is thereby defined as the capability of the model to distinguish landslide-free from landslide-prone areas (Soeters and Van Westen 1996). The basic idea of model validation is to compare the projected landslide distribution with an independent data set of past landslides (cf. Chung and Fabbri 2003; Remondo et al. 2003). Many different validation methods are currently available to evaluate prediction quality and level of model confidence (e.g., Brenning 2005; Beguería 2006; Frattini et al. 2010; Corominas and Mavrouli 2011). A popular method for model validation is the prediction and success rate from Chung and Fabbri (2003). This validation method was used in the present study as it has the advantage to clearly visualize the goodness of fit and the prediction success (cf. Klose et al. 2014a). Model validation using this method requires splitting the spatial landslide inventory into a modeling and validation set. In order to ensure statistical robustness of the model, 15 % of the recorded landslides were classified to the validation group by random selection, an approach that has already been proposed by Neuhäuser et al. (2012a). Modeling of landslide susceptibility is conducted by using only landslide data stored in the modeling set (cf. Chung and Fabbri 2003).

The first step of the performed model validation using the success and prediction rate was to classify the hazard scale into equal intervals. For each of the 19 considered interval classes, the percentage area has been identified in ArcView GIS, and the obtained data were written to a MS Excel data table. By means of GIS-based tools for raster data statistics or extraction of data (Fig. 4.2), areal extent of defined interval classes or the landslide-related values of $SI(x)$ were determined. Subsequent data export to MS Excel enabled categorization of recorded landslides to the different interval classes. Besides percentage area, the percentage share of landslides in each interval class was considered in the validation, and the success or prediction rate resulted from cumulating percentage area and percentage share of landslides in the two data sets. For better data analysis and interpretation, both rates were plotted as curves of their cumulative distribution function, with the result that model accuracy and predictive power is illustrated by the visualized rate curves (cf. Klose et al. 2014a).

Model plausibility can be proved to some extent by testing conditional independence of its input data. A variety of statistical tests has been developed to check conditional independence, including contingency statistics and the χ^2-test (e.g., Agterberg and Cheng 2002; Thiery et al. 2007; Pereira et al. 2012). These two test methods are frequently used in landslide susceptibility modeling (e.g., Lee et al. 2002; Neuhäuser and Terhorst 2007; Regmi et al. 2010) and were therefore applied in this study as well (cf. Klose et al. 2014a). Basic assumption of every statistical landslide susceptibility model is the conditional independence of its input data (Van Westen 2000). This assumption is generally violated when different predisposing factors show comparable spatial patterns. In case of factor dependency, differentiation of the individual factor influence on past landslide occurrence is hampered, wherefore landslide susceptibility is overestimated in areas where dependent predisposing factors spatially coincide (cf. Agterberg and Cheng 2002; Pereira et al. 2012).

4.3 Landslide Cost Assessment Model

4.3.1 Introduction

4.3.1.1 General Overview

Estimation of landslide costs is a crucial but challenging task of special importance for landslide risk assessment. Cost survey (ex-post) and risk analysis (ex-ante) as the two main approaches of landslide loss assessment (e.g., Hallegatte and Przyluski 2010; Meyer et al. 2013; Kreibich et al. 2014) are unable to cope with some characteristic features of landslide impact, including the complex distribution of landslides in space and time (cf. Sect. 1.1) and the problem to reason from landslide intensity and the value and vulnerability of elements at risk to potential costs (i.e., risk estimates ≠ potential costs, vulnerability > 1; cf. Remondo et al. 2008). As a result of strongly variable landslide distribution patterns and the complexity in landslide process mechanisms, the identification, tracking, and documentation of past landslide losses is difficult, which constitutes a crucial problem for cost surveys (see also Highland 2006). Alternatively, case studies show that the costs of landslide damage are rather dependent on the type of landslide repair or mitigation than on the value at risk (e.g., Cornforth 2005; Hearn et al. 2011; Highland 2012; Klose et al. 2012b), as assumed in quantitative risk analysis (e.g., Lee and Jones 2004). To improve availability and reliability of landslide loss data, the existing methods of cost assessment need to be optimized to these characteristic features of landslide impact, but this is poorly realized so far (cf. Klose et al. 2014b).

A literature review has revealed that the state of research in landslide loss studies is characterized by two main eras in which the focus was either on past or future losses (Fig. 1.6). The U.S. thereby shows the longest research history in assessing landslide losses, with the first national cost estimate for the U.S. having been reported by Smith (1958). Most of the many U.S. studies from before the year 2000 were ex-post assessments of which a summary is given by Fleming and Taylor (1980), Schuster and Fleming (1986), Brabb (1989), and Schuster (1996). Since a couple of years, landslide cost assessment experiences a shift from ex-post to ex-ante, which manifests in an increasing number of studies in the field of risk analysis (see below). Besides this recent trend, which started in the mid-1990s, few additional studies, however, have expanded the available toolset for landslide cost assessment by further methods, including socioeconomic evaluation (e.g., Burke et al. 2002; MacLeod et al. 2005), methods of real estate appraisal (cf. Vranken et al. 2013), and probabilistic or database-driven cost modeling (e.g., Crovelli and Coe 2009; Klose et al. 2012b). Even though direct losses received greater attention in the past, indirect ones have not been fully ignored, especially in most recent research (e.g., MacLeod et al. 2005; Zêzere et al. 2007; Ohara et al. 2008). High priority today is placed on the development of strategies for systematic compilation of loss data. Topics such as information content of available data sources, assessment of data accessibility or quality, and design of concepts for efficient data retrieval are of particular interest in this context (e.g., Ashland 2003; Highland 2006; Battistini et al. 2013; Damm and Klose 2014, 2015).

Both approaches of landslide cost estimation benefit from the recent progress in database or web technology and the continuously improving functionality of tools such as GIS. The advancement in digital and/or web-based data organization and archiving over the past decade expanded the pool of available damage and loss data to a large extent and made it much easier to retrieve these data sets for their storage in scientific landslide databases (e.g., Alexander 2008; Battistini et al. 2013). Together with improved methods of computerized data acquisition and data mining, these tools provide valuable support for cost survey, enabling systematic collection of loss data over broad areas. Additionally, new geospatial tools and web resources foster landslide news tracking and disaster documentation, which significantly improves the power and quality of landslide databases used for cost estimation today (Klose et al. 2014c; see also Sect. 2.2.1).

4.3.1.2 Ex-post Assessments

A large part of previous loss studies was focused on ex-post assessment of direct landslide costs for transportation infrastructures, especially the U.S. state highway systems (cf. Chassie and Goughnour 1976; Walkinshaw 1992; Wang et al. 2002; Wyoming Homeland Security 2011; Highland 2012; USGS 2013). From a global perspective, there are only few additional studies dealing with retrieval and analysis of past landslide losses for traffic facilities, and the focus of these studies is also primarily on roads or highways (e.g., Hearn et al. 2008; Public Works Department Malaysia 2008; Negi et al. 2013). The most common method in such transportation-related studies is still cost survey, including expert interviews, questionnaire surveys, and archive studies. This also holds true for much of today's cross-sector studies that are not restricted to a certain type of infrastructure (e.g., Crovelli and Coe 2009; Rahman et al. 2011; Klose et al. 2012b; Vranken et al. 2013; see also Klose et al. 2014b).

Many of the U.S. loss studies from before the year 2000 had a focus on compilation of landslide losses in local or regional case study areas for periods of increased landslide activity such as rainstorm events (e.g., Taylor and Brabb 1972; Shearer et al. 1983; Creasey 1988; Godt 1999). On the basis of cost surveys, these studies developed reference costs at city or county level, and some studies also provided statewide or national cost estimates by extrapolating the obtained local or regional losses (e.g., Schuster 1978; Fleming and Taylor 1980; Brabb 1984). The techniques of cost extrapolation were often very rudimentary, with lack of systematic modeling approaches and validation tools (cf. Klose et al. 2014c). A promising method worth to mention, however, is that of Krohn and Slosson (1976) who projected landslide losses by combining a figure on the number of U.S. citizens living in landslide-prone area with data on landslide costs per private home from southern California. Alternatively, Mathur (1982) presented an approach of cost extrapolation for landslide damage to roads in India. For spatial extrapolation of losses, this study combined an estimate of landslide costs per km of road with a figure on the total length of the road network in several landslide-prone Indian states (see also Klose et al. 2014b).

Despite being widely qualitative and regardless of their deficits in validation and data documentation (cf. Highland 2006), above studies or related global reports (e.g., Brabb and Harrod 1989; Schuster 1996; Schuster and Highland 2001) are still among the most comprehensive ones, wherefore continuing to play a vital role in current landslide research (e.g., Sidle and Ochiai 2006; Kjekstad and Highland 2009; Alimohammadlou et al. 2013). At the European level, by contrast, there are still very few ex post assessments of landslide losses available. Besides the data on landslide losses published in the studies from Klose et al. (2012b) and Vranken et al. (2013), damage and cost information are also included in different geohazard databases, especially those specialized on covering flood and landslide impacts (e.g., Hilker et al. 2009; Trezzini et al. 2013; Damm and Klose 2014, 2015). Previous research on landslide losses in Germany using methods from ex-post assessment or cost extrapolation has been widely restricted to the studies from Krauter (1992), Wolterstorff (2002), and Klose et al. (2012b).

4.3.1.3 Ex-ante Assessments

Ex-ante assessment of landslide losses by means of risk analysis is usually based on a quantitative approach. In quantitative risk analysis (QRA), GIS-based methods are used to calculate landslide risk as function of hazard, vulnerability, and element at risk. Landslide risk is thereby understood as the probability of loss of life or property as consequence of damages caused by potential landslides with given magnitudes and frequencies (cf. Varnes and IAEG 1984; Dai et al. 2002; Lee and Jones 2004; Fell et al. 2005; Van Westen et al. 2006; Corominas et al. 2014). The following risk equation applies for QRA when considering physical infrastructures (roads, buildings, lifelines, etc.) as elements at risk (modified after Dai et al. 2002):

$$R = H(L) \cdot V(E) \cdot E \tag{4.15}$$

with

$$H(L) = P(L) \cdot P(S|L) \tag{4.16}$$

where R is the risk (potential economic loss), $H(L)$ is the landslide hazard, $V(E)$ is the vulnerability of the element at risk, E is the element at risk (monetary value), $P(L)$ is the temporal landslide probability, and $P(S \mid L)$ is the probability of spatial landslide impact (landslide susceptibility).

Most studies of potential landslide losses consider either distributed or site-specific landslide risk. Analysis of distributed landslide risk implies modeling of potential landslide losses for each mapping unit throughout an entire region or case study area (cf. Dai et al. 2002). Examples of recent studies dealing with analysis of distributed landslide risk for different types of physical infrastructures are Blöchl and Braun (2005), Catani et al. (2005), Remondo et al. (2005), Zêzere et al. (2008), Bonachea et al. (2009), Jaiswal et al. (2011), and Erener and Düzgün (2013). In landslide risk assessment for the transportation sector, a major role is played by analyzing the vehicle risk or the risk of loss of life (e.g., Ko Ko et al. 2005; Dorren et al. 2009; Li et al. 2009; Ferlisi et al. 2012; Michoud et al. 2012), whereas potential property losses received little scientific attention so far (e.g., Zêzere et al. 2007; Jaiswal et al. 2010; Bründl et al. 2012). Besides the above studies with a broader geographic perspective, there is also a large number of local or site-specific risk assessments, with either cross-sectoral focus or thematic priority on traffic facilities (e.g., Budetta 2002, 2004; Bell and Glade 2004; Corominas et al. 2005; Sterlacchini et al. 2007; Mousavi et al. 2011; Klimeš and Blahůt 2012).

The workflow of QRA generally involves the following main steps (cf. Australian Geomechanics Society 2000; Crozier and Glade 2005; Fell et al. 2005): (i) hazard identification, (ii) hazard analysis, (iii) consequence analysis, and (iv) risk calculation. Primary goal of the first two steps is to specify landslide hazard as function of landslide susceptibility and temporal landslide probability. While landslide susceptibility is usually determined by using one of the statistical methods presented in Sect. 4.2.1, the calculation of temporal landslide probability is either scenario-based or data-driven (cf. Raetzo et al. 2002; Picarelli et al. 2005; Corominas and Moya 2008; Corominas et al. 2014). Studies such as Dorren et al. (2009) and Sterlacchini et al. (2007) apply the hazard scenarios of BUWAL (1999) that contain predefined landslide return periods. A different strategy to determine temporal landslide probability is the data-driven approach, including frequency analysis of landslide occurrence or triggering events (e.g., Coe et al. 2004; Zêzere et al. 2004; Guzzetti et al. 2005). If considered in risk assessment, landslide magnitude is often approximated using simple scenarios or classes of landslide volume (e.g., Bell and Glade 2004; Remondo et al. 2005; Jaiswal et al. 2010), although different magnitude parameters or frequency-size relations have already been mentioned or applied in hazard assessment (cf. Ojeda-Moncayo et al. 2004; Guzzetti 2005).

The focus of consequence analysis in QRA is on specifying the vulnerability of considered elements at risk (e.g., Glade 2003; Uzielli et al. 2008). Vulnerability is generally understood as the level of potential damage to an element at risk impacted by a landslide of a given magnitude. The level of potential damage is thereby measured on a scale ranging from 0 (no damage) to 1 (total loss) (Varnes and IAEG 1984). Most studies derive physical vulnerability by means of vulnerability functions or indexes that describe the relationship between landslide magnitude and damage based on historical records and expert knowledge or kinematic intensity models (e.g., Glade 2003; Galli and Guzzetti 2007; Papathoma-Köhle et al. 2011; Pitilakis et al. 2011; Silva and Pereira 2014). The last step of QRA refers to the monetization of elements at risk, whereby either market values or reconstruction costs are used for monetization (cf. Alexander 2005). Landslide risk in QRA is often calculated as percentage of the market values of elements at risk (e.g., Blöchl and Braun 2005; Bonachea et al. 2009), and as these market values hardly reflect direct damage costs (Klose et al. 2014c), QRA is usually fraught with significant uncertainty. This is why an ex-post approach has been chosen in this study to assess landslide losses for highways in the Lower Saxon Uplands (cf. Klose et al. 2014b).

4.3.2 Model Description

4.3.2.1 General Overview

The developed method provides a general framework and toolset for estimation and regionalization of direct landslide costs for transportation infrastructures (cf. Klose et al. 2014b). To apply the tools provided by this method in a regional cost assessment, it is necessary to specify and customize them according to the regional data situation and the sociotechnical conditions of the study area. In this study, the method is designed to support regional cost estimation for highways in study areas characterized as follows: (i) high level of societal risk aversion, (ii) highly developed highway systems, and (iii) advanced coping capacity in both technological and financial terms. The configuration of the different tools is based on management practices, mitigation concepts, and cost data from transportation planning, highway maintenance, and engineering in the Central Uplands of Germany.

The workflow architecture of this method is defined by a bottom-up approach of cost estimation that pursues the goal to spatially extrapolate past and current landslide losses from a local case study area to regional level (Fig. 4.4). A reliable time period for cost compilation on a local scale is about the previous 20–30 years. The reason for this reference period is to consider cost volatility caused by fluctuating landslide activity. A reference period of 20–30 years before the present was also proposed by Walkinshaw (1992). Such an ex-post approach is adapted to the spatiotemporal characteristics of landslide impact and increases the reliability of regional studies. The method is composed of two tiers of cost estimation: one on a local level, and one on a regional level (Fig. 4.4). Tier 1 provides tools for local cost compilation, and tier 2 is focused on regional cost extrapolation. Both tiers are linked by tools that enable data fusion (cf. Klose et al. 2014b).

The basis and starting point of cost estimation in this method is a landslide database, for example, the one available for this research (Sect. 2.2). Most landslide databases store data sets on landslide location and thus provide functionality for landslide susceptibility modeling (cf. Sect. 2.1). This is important as a regional landslide susceptibility model is at the heart of this methodology, fulfilling two main tasks: (i) decision support for the selection of a representative case study area for local cost compilation, and (ii) provision of tools for the regionalization of local landslide losses. In addition to data sets on landslide location, some landslide databases, as shown in Sect. 2.1, also contain information on landslide impact and types of landslide damage. These data sets are vital for cost compilation in the

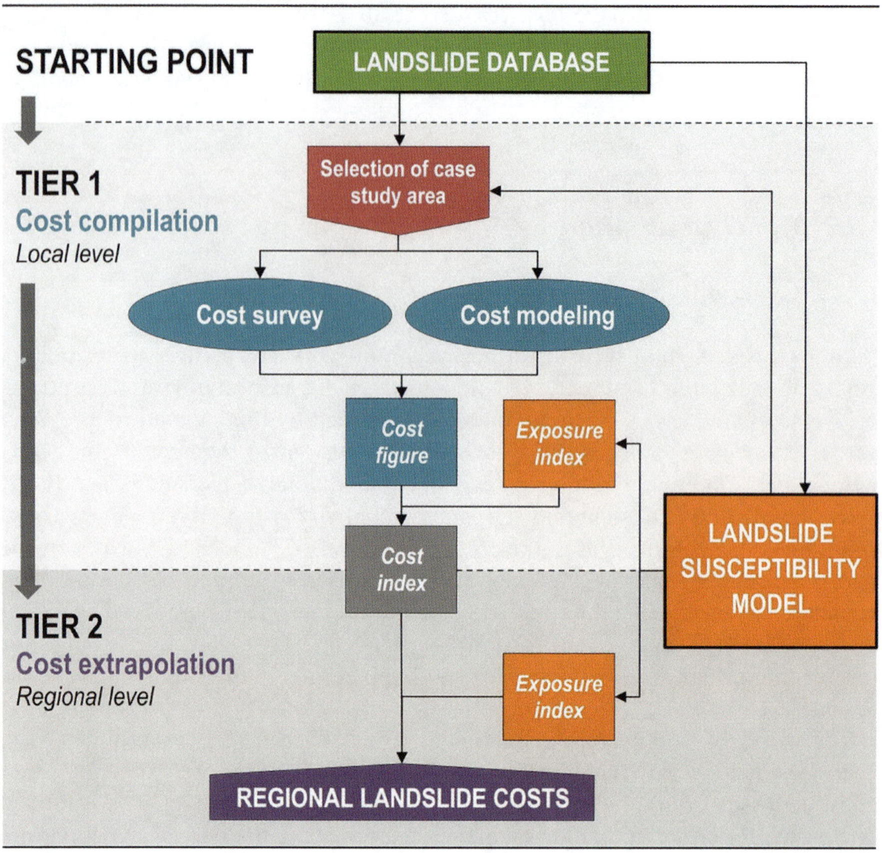

Fig. 4.4 General framework and tools of the method for landslide cost modeling. The method is based on a bottom-up approach and combines tools for local cost compilation with a landslide susceptibility model for the regionalization of local landslide losses. A key role in cost extrapolation is played by a local cost index and a local or regional exposure index (*Source* Klose et al. 2014b)

first tier of this methodology. The objective of local cost compilation is to create a complete and consistent loss record for a relevant case study area by the application of various techniques to broaden and monetize data sets stored in an underlying landslide database. In this context, the two most important tools are cost survey and cost modeling. The main idea of cost survey is to retrieve first-hand loss data for recent landslides through targeted data mining of official accounting and archive systems. Alternatively, cost modeling is primarily intended for cost estimation of landslide damage older than the last 10–15 years, as this is the time period when official accounting and archive systems are electronically available and provide the most detailed loss data. The basic principle of cost modeling is to classify the landslide damage events documented in the landslide database to the following modeling concepts: (i) landslide disaster management process models (LDMM) that provide the costs involved in coping with landslide hazards over the full disaster cycle, and (ii) cost categories of the total costs of certain types of landslide damage. The result of local cost compilation is a cost figure that gives the annual average costs per kilometer of highway in the case study area (cf. Klose et al. 2014b).

In the second tier, the regionalization of local landslide losses is realized by GIS-based cost extrapolation using a landslide susceptibility model. The purposes of the landslide susceptibility model in cost extrapolation are as follows: (i) to identify sections of the highway network exposed to landslide hazards, (ii) to support the development of a local and regional exposure index, and (iii) to update the local cost figure to a cost index. These indexes show the capacity for cost extrapolation because supporting data fusion to realize the connection between the two tiers of this methodology. The basic assumption behind the concept of cost extrapolation is that hazard areas on a regional level probably experience similar annual costs per kilometer of highway as comparable areas at risk on a local level. This enables cost extrapolation to a regional level by simple operations on the basis of the local cost index and the regional exposure index (cf. Klose et al. 2014b)

4.3.2.2 Landslide Database and Study Areas

The landslide database used in the loss study for Lower Saxon highways within this research is the same than that applied for landslide susceptibility modeling (Sect. 4.2.2). This combined database thus includes the developed spatial landslide inventory and the regional database subset for Lower Saxony of the available landslide database. The main function of these two information sources in this loss study is to provide a data pool for cost modeling and to enable the application of the developed landslide susceptibility model. With regard to database-driven cost modeling, the regional database subset contains several essential data tables (cf. Sect. 2.2.2), including that on landslide process mechanism (movement type, size, magnitude, etc.), landslide impact (damage profile, first response, etc.), and landslide mitigation (repair or prevention measures). This database subset stores a consistent record of such impact-related data sets for 33 landslides at highways in the Upper Weser area (Fig. 3.1) between 1980 and 2010. The high availability of detailed landslide data for the Upper Weser area in this reference period justified choosing this regionally representative area as case study area for local cost compilation (cf. Klose et al. 2014b). Based on the loss data compiled for highways in the Upper Weser area, the direct landslide costs affecting the highway system in the Lower Saxon Uplands (core area, cf. Fig. 3.1) have been regionally extrapolated. For further information on the landslide susceptibility model for cost extrapolation, it is referred to the explanations on model input data and development given in Sect. 4.2.2.

4.3.2.3 Local Cost Compilation

Cost Survey

A cost survey was conducted at the Regional Office Gandersheim of the Lower Saxony Department of Transportation (NLStBV). This cost survey enabled the acquisition of data on landslide costs for highways in the Upper Weser area between 2001 and 2010. The obtained loss data exclusively refer to landslides that are already recorded in the landslide database. The costs of these most recent landslides are well-documented in the Project Information and Management System (PRIMAS) operated by the NLStBV Regional Office Gandersheim. This system is a MS Excel database for accounting and controlling of the business processes during one fiscal year. PRIMAS records for construction and maintenance projects with costs of more than US$70,000 all payment transactions, internal labor costs, and a brief specification of services. The financing of landslide repair below these minimum costs is based on fixed maintenance budgets that hamper cost itemization. PRIMAS cost data are used for the evaluation of seven (21 %) major projects of landslide repair and mitigation. A key advantage of cost survey based on PRIMAS is that this system provides a large part of the actual landslide costs. Only costs associated with road closure are often included in maintenance budgets and must be evaluated by cost modeling. This is also the case for the costs of first response that are usually ignored in PRIMAS. The only prerequisite for data retrieval in PRIMAS is the difficult task of identifying the right project ID for the relevant landslide damage event (cf. Klose et al. 2014b).

Cost Modeling

Landslide disaster management process models (LDMM) are used for database-driven cost modeling of recorded landslide damage to highways before the year 2001. By contrast, cost categorization is only presented in the discussion as an alternative modeling approach with reduced data requirements. The necessary landslide database information for cost modeling based on LDMMs include a landslide process description, a damage profile, and a fact sheet of repair or mitigation. The two basic assumptions of cost modeling are as follows: (i) highways in the Lower Saxon Uplands are often affected by similar types of landslides and landslide damage, and (ii) landslide disaster management for highways in this region usually follows a standardized response, recovery, and/or mitigation management process. Both assumptions are empirically verified and constitute a necessary precondition for cost modeling by allowing the monetization of landslide damage with a high degree of standardization (cf. Klose et al. 2014b).

Three different LDMMs were designed to model landslide losses for highways: (i) recovery process model, (ii) mitigation process model, and (iii) maintenance process model. Each LDMM covers one key segment of the disaster cycle and simulates the cost-relevant steps involved in the process of coping with or preventing landslide damage. The LDMMs constitute cost chains that display the major cost factors of disaster management (first response, road closure, etc.) and fulfill the function to provide the basic framework for cost modeling based on landslide databases. Flowcharts are used as a modeling technique to visualize the disaster management processes for exemplary types of landslide damage to highways in the Lower Saxon Uplands. The development of the LDMMs is based on qualitative data material collected by conducting expert interviews with personnel of emergency management agencies (police and fire departments; Federal Agency of Technical Relief, THW) and the NLStBV Regional Office Gandersheim. Further information sources of high importance are federal and/or state emergency laws (Nds. SOG, NRettDG), disaster response laws (NKatSG, THW-Gesetz), and road construction law (NStrG).

The workflow of cost modeling starts with the classification of a landslide damage event (i.e., database entry) to a LDMM (Fig. 4.5). By means of the LDMM, the cost factors of this landslide damage event are determined, which is the first step of cost modeling. Subsequently, cost modules are used to monetize the identified cost factors. Cost modules refer to specific emergency, repair, mitigation or maintenance measures and provide an estimate of their total costs. A distinction is made between basic and complex cost modules. Basic cost modules (e.g., geotechnical report) are based on a fixed cost rate and thus are easy to calculate. The costing of complex cost modules, however, is more sophisticated. Thus, one part of the complex cost modules is based on estimates of average costs per meter or square meter (e.g., catch fence, rockfall drapery), while the other part (e.g., road closure, rock buttress) relies on cost formulas that require entering basic or process-related data (e.g., road closure time; depth, length, width of slip surface). Complex cost modules for repair or mitigation structures calculated using average costs are mostly differentiated in categories of usual sizes, for example, catch fence with low, medium or high energy absorption capacity. The costing by means of LDMMs is done incrementally, with each cost module being individually calculated. In the last step, the monetized cost factors are totaled, so as to obtain the overall costs of the landslide damage event. This process of cost modeling is repeated for all database entries (cf. Klose et al. 2014b).

The price data integrated in cost modules are extracted from representative construction cost databases (Baupreislexikon, sirAdos) or refer to cost proxies gathered by questionnaire surveys (mail surveys, $n = 25$) and expert interviews (face-to-face and telephone surveys, $n = 50$) at geotechnical engineering companies, emergency agencies, and the NLStBV. All prices of these data sources reflect current market prices, and most of them are net prices, which is why they are subject to 19 % value added tax (standard VAT rate). The two exemptions are internal labor costs and fees for emergency services. The design of complex cost modules is based on geotechnical concepts and dimensioning rules found in literature or obtained from engineering practice. Alternatively, directives and guidelines for road design (RAS-Q, RAS-Ew, RAS-LG, M Geok E, etc.) and traffic control (RSA-95, RUB-92) are a valuable basis for the development of complex cost modules. The tabulation and costing of landslide losses compiled by cost survey and

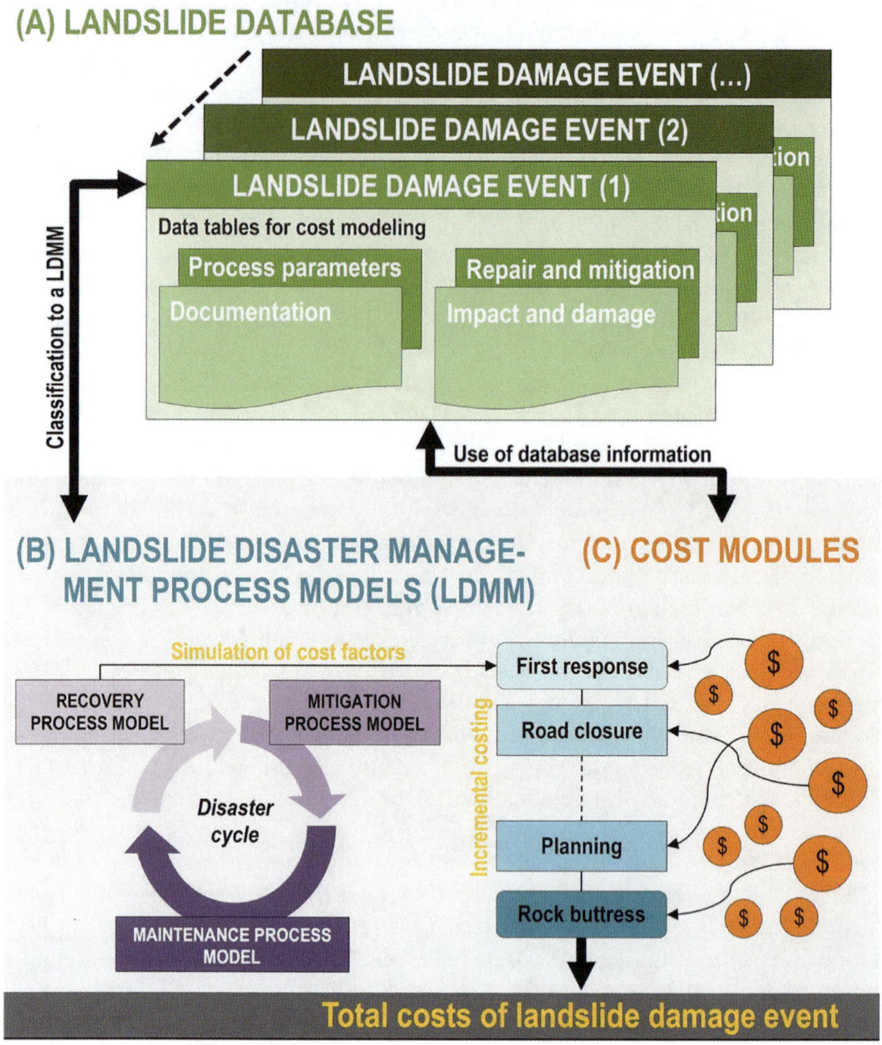

Fig. 4.5 Workflow of database-driven landslide cost modeling with the main tools to monetize damage events stored in a landslide database. The procedure uses landslide disaster management process models (LDMM) and cost modules for simulation and costing of the cost-relevant steps in coping with or preventing landslide damage over the full disaster cycle (*Source* Klose et al. 2014b)

cost modeling are performed using a MS Excel application. This application is a toolset, including (i) a data table giving the total costs on local and regional level, (ii) a tool for cost modeling integrated in the data form for each landslide damage event, and (iii) data tables storing the price data and the data for cost extrapolation.

The different data tables are linked by relationships, and data processing uses custom functions, i.e., cost formulas for certain cost modules (e.g., embankment infill buttress), both of which enable semiautomatic costing. For example, by entering the input data of a cost module (e.g., length and height of embankment) in the data form of the landslide damage event, this application writes the costs of this cost module in the data table of cost tabulation.

The result of cost compilation is a cost figure for highways in the Upper Weser area. This cost figure is based on the annual average of the total losses over the time period 1980–2010. The main idea underlying the cost figure is to break down the annual average costs on the total length of the highway network at local level. Consequently, the cost figure specifies annual average costs per kilometer of highway in the Upper Weser area (cf. Klose et al. 2014b).

4.3.2.4 Example of Cost Modeling

The concept of cost modeling is illustrated using the example of a shallow landslide blocking a highway in the Upper Weser area in the year 1994 (Fig. 4.6). First, the landslide damage event is classified to the right LDMM, which is the recovery process model. According to this LDMM, disaster management starts with first response by a police patrol and a basic firefighting unit. The emergency responders report the landslide damage to the local highway maintenance depot that closes the road and installs a detour. Afterwards, engineers of the NLStBV Regional Office Gandersheim conduct an on-site inspection to define further actions. The first step usually is debris removal undertaken by contracted construction companies. The primary goal is to reopen the highway to single-lane traffic as soon as possible. Once the traffic is moving again, the planning of landslide repair is made. The planners generally consult expert opinion and rely on a geotechnical report. Landslide repair starts only after public awarding of the construction project. The realization of the rock buttress completes the recovery process.

The cost table of this landslide damage event (Fig. 4.6) lists the applied cost modules with their input data and costs. The fixed rates of the cost module first response and on-site inspection are based on official cost rates and benchmarks of operation time, number of personnel, and equipment. An orthogonal highway network is assumed to estimate the costs of the road closure. This cost module is designed on the basis of a traffic control plan for road closure with off-site detour and cost rates from traffic control companies. While the price of the cost module geotechnical report constitutes a standard market value, the cost module planning and building site equipment are calculated as a percentage of the net construction costs. Landslide process parameters and various assumptions concerning labor, machinery, and performance enable to apply the cost module debris and vegetation removal. Besides cost rates for necessary traffic signs, the cost module traffic control is based on an exemplary traffic control plan (lane closure on two-lane road using traffic signals). As is the case with the cost module road closure, the

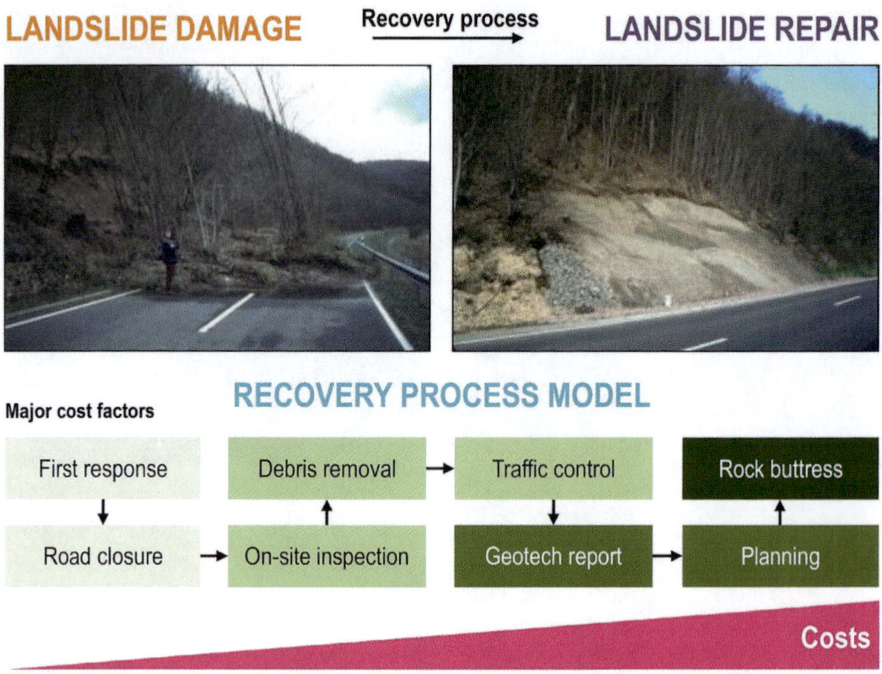

COST TABLE

COST MODULE	INPUT DATA	COSTS [in USD thousand]
First response	No, fixed rate	4.5
Road closure*	Time: t [day]	8.6
On-site inspection	No, fixed rate	0.5
Debris removal	Landslide volume: v [m³]	14.4
+ Vegetation removal	Slip surface: length l [m], width w [m]	9.2
Traffic control	Time [days]	2.8
Geotechnical report	No, fixed rate	2.0
Planning	No, percentage rate	4.9
Building site equipment	No, percentage rate	1.0
Rock buttress	Slip surface: depth d [m], length l [m], width w [m]	15.8
	TOTAL	65

Example of cost formula

*7,800 + 200 × t, where t = 4

Fig. 4.6 Example of cost modeling for a landslide damage event affecting a highway near the city of Hann. Münden (Upper Weser area) in the year 1994. Based on the developed recovery process model and relevant cost modules, the costs of recovery ranging from first response to final landslide repair are identified and calculated using input data from the available landslide database (*Source* Klose et al. 2014b)

time of traffic control is derived from the general landslide documentation in the database. The costing of the rock buttress uses a standard repair concept and database information on the depth, width, and length of the slip surface. The total costs of this landslide damage event are thus estimated at about US$65,000 (cf. Klose et al. 2014b).

4.3.2.5 Regional Cost Extrapolation: Exposure Indexes, Cost Index, and Validation

The information on landslide susceptibility obtained by applying the bivariate statistical modeling approach presented in Sect. 4.2.2 is used to calculate a local and regional exposure index. This index measures how many kilometers of highway are located in potential landslide hazard area on local or regional level. The calculation of the exposure index is based on spatial intersection of the hazard area identified by landslide susceptibility modeling with a local or regional data set of the location of highways. Furthermore, the information on landslide susceptibility is used to update the cost figure to a cost index. This is done by replacing the reference base of the cost figure by the local exposure index. Consequently, the cost index for the Upper Weser area specifies the costs per kilometer of highway at risk of landslides. The total annual average costs for highways in the Lower Saxon Uplands are obtained by multiplying the local cost index and the regional exposure index (cf. Klose et al. 2014b).

The validation of the obtained cost estimates is an essential part of this method and concerns both cost modeling and cost extrapolation. A validation system enabling comparison of the results of cost modeling with reference data from cost survey provides options to check data accuracy and is therefore developed and applied in this study. Many different techniques are available to validate landslide susceptibility models (cf. Sect. 4.2.2), whereby the prediction and success rate from Chung and Fabbri (2003) is also suitable for verifying the procedure and plausibility of the developed approach for cost extrapolation. The applied validation methods together with the validated results of the loss study for Lower Saxon highways are presented in Sects. 5.3.2 and 5.3.3 in detail.

References

Adrien, N. G. (2004). *Computational hydraulics and hydrology: An illustrated dictionary*. Boca Raton: CRC Press. 452 pp.

AG BODEN. (2005). *Bodenkundliche Kartieranleitung*. Hannover: Schweizerbart. 438 S.

Agterberg, F. P., & Cheng, Q. (2002). Conditional independence test for weights-of-evidence modeling. *Natural Resources Research, 11*(4), 249–255.

Aleotti, P., & Chowdhury, R. (1999). Landslide hazard assessment: Summary review and new perspectives. *Bulletin of Engineering Geology and the Environment, 58*, 21–44.

Alexander, D. (1992). On the causes of landslides: Human activities, perception, and natural processes. *Environmental Geology and Water Sciences, 20*(3), 165–179.

Alexander, D. (2005). Vulnerability to landslides. In T. Glade, M. G. Anderson, & M. J. Crozier (Eds.), *Landslide hazard and risk* (pp. 175–198). Chichester: Wiley.

Alexander, D. (2008). A brief survey of GIS in mass-movement studies, with reflections on theory and methods. *Geomorphology, 94*, 261–267.

Alimohammadlou, Y., Najafi, A., & Yalcin, A. (2013). Landslide process and impacts: A proposed classification method. *Catena, 104*, 219–232.

Arbeitskreis Grundwasserneubildung FH-DGG. (1977). Methoden zur Bestimmung der Grundwasserneubildungsrate. *Geologisches Jahrbuch C, 19*, 3–98.

Ashland, F. X. (2003). *The feasibility of collecting accurate landslide-loss data in Utah* (25 pp). Open-File Report 410. Salt Lake City: Utah Geological Survey.

Australian Geomechanics Society. (2000). Landslide risk management concepts and guidelines. *Australian Geomechanics, 35*(1), 49–92.

Ayalew, L., & Yamagishi, H. (2005). The application of GIS-based logistic regression for landslide susceptibility mapping in the Kakuda-Yahiko Mountains, Central Japan. *Geomorphology, 65*, 15–31.

Ayalew, L., Yamagishi, H., & Ugawa, N. (2004). Landslide susceptibility mapping using GIS-based weighted linear combination, the case in Tsugawa area of Agano River, Niigata Prefecture, Japan. *Landslides, 1*, 73–81.

Battistini, A., Segoni, S., Manzo, G., Catani, F., & Casagli, N. (2013). Web data mining for automatic inventory of geohazards at national scale. *Applied Geography, 43*, 147–158.

Beguería, S. (2006). Validation and evaluation of predictive models in hazard assessment and risk management. *Natural Hazards, 37*, 315–329.

Bell, R., & Glade, T. (2004). Quantitative risk analysis for landslides – Examples from Bíldudalur, NW-Iceland. *Natural Hazards and Earth System Sciences, 4*, 117–131.

Bell, R., Mayer, J., Pohl, J., Greiving, S., & Glade, T. (2010). *Integrative Frühwarnsysteme für gravitative Massenbewegungen (ILEWS). Monitoring, Modellierung, Implemenierung*. Essen: Klartext Verlag. 270 S.

Bíl, M., & Müller, I. (2008). The origin of shallow landslides in Moravia (Czech Republic) in the spring of 2006. *Geomorphology, 99*, 246–253.

Blaney, H. F., & Criddle, W. D. (1962). *Determining consumptive use and irrigation water requirements* (59 pp). Technical Bulletin Nr. 1275. Washington D.C.: Agricultural Research Service.

Blöchl, A., & Braun, B. (2005). Economic assessment of landslide risks in the Swabian Alb, Germany – Research framework and first results of homeowners' and experts' survey. *Natural Hazards and Earth System Sciences, 5*, 389–396.

Bonachea, J., Remondo, J., Díaz de Terán, J. R., González-Díez, A., & Cendrero, A. (2009). Landslide risk models for decision making. *Risk Analysis, 29*(11), 1629–1643.

Bonham-Carter, G. F., Agterberg, F. P., & Wright, D. F. (1989). Weights of evidence modeling: A new approach to mapping mineral potential. In F. P. Agterberg, & G. F. Bonham-Carter (Eds.), *Statistical Applications in the Earth Sciences 89-9* (pp. 171–183). Geological Survey of Canada.

Boogard, T. A., & Van Asch, T. W. J. (2002). The role of the soil moisture balance in the unsaturated zone on movement and stability of the Beline landslide, France. *Earth Surface Processes and Landforms, 27*, 1177–1188.

Bormann, H., Breuer, L., Giertz, S., Huisman, J. A., & Viney, N. R. (2009). Spatially explicit versus lumped models in catchment hydrology – Experiences from two case studies. In P. C. Baveye, J. Mysiak, & M. Laba (Eds.), *Uncertainties in environmental modelling and consequences for policy making* (pp. 3–26). Berlin: Springer.

Brabb, E. E. (1984). *Minimum landslide damage in the United States, 1973–1983* (4 pp). Open-File Report 84–486. Reston: U.S. Geological Survey.

Brabb, E. E. (1989). Landslides: Extent and economic significance in the United States. In E. E. Brabb, & B. L. Harrod (Eds.), *Landslides: Extent and economic significance* (pp. 25–50). Rotterdam: A.A. Balkema.

Brabb, E. E. (1991). The world landslide problem. *Episodes, 14*(1), 52–61.

Brabb, E. E., & Harrod, B. L. (Eds.). (1989). *Landslides: Extent and economic significance.* Rotterdam: A.A. Balkema. 385 pp.

Brechtel, H. M., & Pavlov, M. B. (1977). *Niederschlagsbilanz von Waldbeständen verschiedener Baumarten und Altersklassen in der Rhein-Main-Ebene – Schätztabellen – (Arbeitspapier).* Arbeitsausschuss Nutzung und Erhaltung der Kulturlandschaft, Arbeitsgruppe Wasserwirtschaft in der Kultur- und Erholungslandschaft.

Brenning, A. (2005). Spatial prediction models for landslide hazards: Review, comparison and evaluation. *Natural Hazards and Earth System Sciences, 5*, 853–862.

Brocca, L., Melone, F., & Moramarco, T. (2008). On the estimation of antecedent wetness conditions in rainfall-runoff modeling. *Hydrological Processes, 22*, 629–642.

Brocca, L., Ponziani, F., Moramarco, T., Melone, F., Berni, N., & Wagner, W. (2012). Improving landslide forecasting using ASCAT-derived soil moisture data: A case study of the Torgiovannetto landslide in central Italy. *Remote Sensing, 4*, 1232–1244.

Bründl, M., Winkler, C., & Baumann, R. (2012). "Econome Railway" – A new calculation method and tool for comparing the effectiveness and the cost-efficiency of protective measures along railways. *Interpraevent, 12*, 933–943.

Budetta, P. (2002). Risk assessment from debris flows in pyroclastic deposits along a motorway, Italy. *Bulletin of Engineering Geology and the Environment, 61*(4), 293–301.

Budetta, P. (2004). Assessment of rockfall risk along roads. *Natural Hazards and Earth System Sciences, 4*, 71–81.

Bulut, F., Boynukalin, S., Tarhan, F., & Ataoglu, E. (2000). Reliability of landslide isopleth maps. *Bulletin of Engineering Geology and the Environment, 58*(2), 95–98.

Burke, T. J., Sattler, D. N., & Terich, T. (2002). The socioeconomic effects of a landslide in Western Washington. *Environmental Hazards, 4*, 129–136.

BUWAL. (1999). *Risikoanalyse bei gravitativen Naturgefahren: Methode* (155 S). Umwelt-Materialien Nr. 107/I Naturgefahren. Bern:Bundesamt für Umwelt, Wald und Landschaft (BUWAL).

Caine, N. (1980). The rainfall intensity-duration control of shallow landslides and debris flows. *Geografiska Annaler, 62 A*, 23–27.

Cardinali, M., Galli, M., Guzzetti, F., Ardizzone, F., Reichenbach, P., & Bartoccini, P. (2006). Rainfall induced landslides in December 2004 in south-western Umbria, central Italy: Types, extent, damage and risk assessment. *Natural Hazards and Earth System Sciences, 6*, 237–260.

Carrara, A., Cardinali, M., Guzzetti, F., & Reichenbach, P. (1995). GIS technology in mapping landslide hazard. In A. Carrara, & F. Guzzetti (Eds.), *Geographical information systems in assessing natural hazards* (pp. 135–175). Dordrecht: Kluwer Academic Publisher.

Carrara, A., Guzzetti, F., Cardinali, M., & Reichenbach, P. (1999). Use of GIS technology in the prediction and monitoring of landslide hazard. *Natural Hazards, 20*, 117–135.

Cascini, L., Bonnard, C., Corominas, J., Jibson, R., & Montero-Olarte, J. (2005). Landslide hazard and risk zoning for urban planning and development. In O. Hungr, R. Fell, R. Couture, & E. Eberhardt (Eds.), *Landslide risk management* (pp. 199–235). London: Taylor & Francis Group.

Catani, F., Casagli, N., Ermini, L., Righini, G., & Menduni, G. (2005). Landslide hazard and risk mapping at catchment scale in the Arno River basin. *Landslides, 2*, 329–342.

Çevik, E., & Topal, T. (2003). GIS-based landslide susceptibility mapping for a problematic segment of the natural gas pipeline, Hendek (Turkey). *Environmental Geology, 44*, 949–962.

Chacón, J., Irigaray, C., Fernández, T., & El Hamdouni, R. (2006). Engineering geology maps: Landslides and geographical information systems. *Bulletin of Engineering Geology and the Environment, 65*, 341–411.

Chassie, R. G., & Goughnour, R. D. (1976). National highway landslide experience. *Highway Focus, 8*, 1–9.

Chleborad, A. F. (2003). *Preliminary evaluation of a precipitation threshold for anticipating the occurrence of landslides in the Seattle, Washington, area.* U.S. Geological Survey Open-File Report 03-463. Retrieved July 28, 2014, from http://pubs.usgs.gov/of/2003/ofr-03-463/ofr-03-0463.html.

Cho, S. E., & Lee, S. R. (2001). Instability of unsaturated soil slopes due to infiltration. *Computers and Geotechnics, 28*, 185–208.

Chung, C.-J. F., & Fabbri, A. G. (2003). Validation of spatial prediction models for landslide hazard mapping. *Natural Hazards, 30*, 451–472.

Coe, J. A., Michael, J. A., Crovelli, R. A., Savage, W. Z., Laprade, W. T., & Nashem, W. D. (2004). Probabilistic assessment of precipitation-triggered landslides using historical records of landslide occurrence, Seattle, Washington. *Environmental & Engineering Geoscience, 10*(2), 103–122.

Collins, B. D., & Znidarcic, D. (2004). Stability analyses of rainfall induced landslides. *Journal of Geotechnical and Geoenvironmental Engineering, 130*(4), 362–372.

Conforti, M., Robustelli, G., Muto, F., & Critelli, S. (2012). Application and validation of bivariate GIS-based landslide susceptibility assessment for the Vitravo river catchment (Calabria, south Italy). *Natural Hazards, 61*, 127–141.

Cornforth, D. H. (2005). *Landslides in practice: Investigations, analysis, and remedial/preventive options in soils.* Hoboken: Wiley. 624 pp.

Corominas, J., & Mavrouli, O. (2011). *Recommended procedures for validating landslide hazard and risk models and maps* (162 pp). SafeLand FP7, Deliverable D2.8. Retrieved August 1, 2014, from http://www.safeland-fp7.eu/results/Documents/D2.8.pdf.

Corominas, J., & Moya, J. (2008). A review of assessing landslide frequency for hazard zoning purposes. *Engineering Geology, 102*, 193–213.

Corominas, J., Copons, R., Moya, J., Vilaplana, J. M., Altimir, J., & Amigó, J. (2005). Quantitative assessment of the residual risk in a rockfall protected area. *Landslides, 2*, 343–357.

Corominas, J., van Westen, C., Frattini, P., Cascini, L., Malet, J.-P., Fotopoulou, S., et al. (2014). Recommendations for the quantitative analysis of landslide risk. *Bulletin of Engineering Geology and the Environment, 73*, 209–263.

Costanzo, D., Rotigliano, E., Irigaray, C., Jiménez-Perálvarez, J. D., & Chacón, J. (2012). Factors selection in landslide susceptibility modelling on large scale following the gis matrix method: Application to the river Beiro basin (Spain). *Natural Hazards and Earth System Sciences, 12*, 327–340.

Creasey, C. L. (1988). Landslide damage: A costly outcome of the storm. In S. D. Ellen, & G. F. Wieczorek (Eds.), *Landslides, floods, and marine effects of the storm of January 3–5, 1982, in the San Francisco Bay Region, California* (pp. 195–203). U.S. Geological Survey Professional Paper 1434. Washington: United States Government Printing Office.

Crosta, G. B., Agliardi, F., Frattini, P., & Sosio, R. (2012). *Landslide triggering mechanisms in Europe – Overview and state of the art* (378 pp). SafeLand FP7, Deliverable 1.1. Retrieved August 1, 2014, from http://www.safeland-fp7.eu/results/Documents/D1.1_revised.pdf.

Crovelli, R. A., & Coe, J. A. (2009). Probabilistic estimation of numbers and costs of future landslides in the San Francisco Bay region. *Georisk, 3*, 206–223.

Crozier, M. J. (1986). *Landslides: Causes, consequences & environment.* London: Croom Helm. 252 pp.

Crozier, M. J. (1999). Prediction of rainfall-triggered landslides: A test of the antecedent water status model. *Earth Surface Processes and Landforms, 24*, 825–833.

Crozier, M. J., & Glade, T. (2005). Landslide hazard and risk: Issues, concepts and approach. In T. Glade, M. G. Anderson, & M. J. Crozier (Eds.), *Landslide hazard and risk* (pp. 1–38). Chichester: Wiley.

Dai, F. C., Lee, C. F., & Ngai, Y. Y. (2002). Landslide risk assessment and management: An overview. *Engineering Geology, 64*, 65–87.

Damm, B. (2002). Rutschungen im Fulda- und Oberweserraum (Nordhessen/Südniedersachsen) – Ursachen, Auslöser und zeitliche Häufungen. In F. Fiedler, F. Nestmann, & M. Kohler (Eds.), *Naturkatastrophen in Mittelgebirgsregionen* (pp. 129–147). Berlin: Verlag für Wissenschaft und Forschung.

Damm, B. (2005). Gravitative Massenbewegungen in Mittelgebirgsräumen – Geowissenschaftlich-historische Analyse von Massenbewegungsprozessen in Nordhessen und Südniedersachsen als Beitrag zur Angewandten Geomorphologie (262 S. + Anhang, unveröffentlicht). Habilitationsschrift, Universität Eichstätt-Ingolstadt.

Damm, B., & Klose, M. (2014). Landslide database for the Federal Republic of Germany: A tool for analysis of mass movement processes and impacts. In K. Sassa, P. Canuti, & Y. Yin (Eds.), *Landslide Science for a Safer Geoenvironment, Vol. 2: Methods of Landslide Studies* (pp. 787–792). Berlin: Springer.

Damm, B., & Klose, M. (2015). The landslide database for Germany: Closing the gap at national level. *Geomorphology*. doi:10.1016/j.geomorph.2015.03.021.

De Blasio, F. V. (2011). *Introduction to the physics of landslides. Lecture notes on the dynamics of mass wasting*. Dordrecht: Springer. 400 pp.

de Jong, R., & Bootsma, A. (1996). Review of recent developments in soil water simulation models. *Canadian Journal of Soil Science, 76*(3), 263–273.

DeGraff, J. V., & Canuti, P. (1988). Using isopleth mapping to evaluate landslide activity in relation to agricultural practices. *Bulletin of the International Association of Engineering Geology, 38*, 61–71.

Dilley, M., Chen, R. S., Deichmann, U., Lerner-Lam, A. L., Arnold, M., Agwe, J., et al. (2005). *Natural disaster hotspots: A global risk analysis*. Washington D.C.: The World Bank. 132 pp.

Dorren, L. K. A., Sandri, A., & Raetzo, H. (2009). Landslide risk mapping for the entire Swiss national road network. In J.-P. Malet, A. Remaître, & T. Boogard (Eds.), *Landslide processes: From geomorphologic mapping to dynamic modelling* (pp. 277–281). Strasbourg: European Centre on Geomorphological Hazards (CERG).

Duncan, J. M., & Wright, S. G. (2005). *Soil strength and slope stability*. Hoboken: Wiley. 312 pp.

Dyck, S. (1983). Overview on the present status of the concepts of water balance models. In A. Van der Beken, & A. Herrmann (Eds.), *New approaches in water balance computations* (pp. 3–19). IAHS Publication No. 148.

Erener, A., & Düzgün, H. B. S. (2013). A regional scale quantitative risk assessment for landslides: Case of Kumluca watershed in Bartin, Turkey. *Landslides, 10*, 55–73.

Ernstberger, H. (1987). *Einfluss der Landnutzung auf Verdunstung und Wasserbilanz. Bestimmung der aktuellen Evapotranspiration von unterschiedlich genutzten Standorten zur Ermittlung der Wasserbilanz von Einzugsgebieten in unteren Mittelgebirgslagen Hessens*. Kirchzarten: Verlag Beiträge zur Hydrologie. 189 S.

Fell, R., Ho, K. K. S., Lacasse, S., & Leroi, E. (2005). A framework for landslide risk assessment and management. In O. Hungr, R. Fell, R. Couture, & E. Eberhardt (Eds.), *Landslide risk management* (pp. 3–26). London: Taylor & Francis Group.

Fell, R., Corominas, J., Bonnard, C., Cascini, L., Leroi, E., & Savage, W. Z., on behalf of the JTC-1 Joint Technical Committee on Landslides and Engineered Slopes (2008). Guidelines for landslide susceptibility, hazard and risk zoning for land use planning. *Engineering Geology, 102*, 85–98.

Ferlisi, S., Cascini, L., Corominas, J., & Matano, F. (2012). Rockfall risk assessment to persons travelling in vehicles along a road: The case study of the Amalfi coastal road (Southern Italy). *Natural Hazards, 62*, 691–721.

Fleming, R. W., & Taylor, F. A. (1980). *Estimating the costs of landslide damage in the United States* (21 pp). Geological Survey Circular 832. Arlington: U.S. Geological Survey.

Frattini, P., Crosta, G., & Sosio, R. (2009). Approaches for defining thresholds and return periods for rainfall-triggered shallow landslides. *Hydrological Processes, 23*, 1444–1460.

Frattini, P., Crosta, G., & Carrara, A. (2010). Techniques for evaluating the performance of landslide susceptibility models. *Engineering Geology, 111*, 67–72.

Fressard, M., Thiery, Y., & Maquaire, O. (2014). Which data for quantitative landslide susceptibility mapping at operational scale? Case study of the Pays d'Auge plateau hillslopes (Normandy, France). *Natural Hazards and Earth System Sciences, 14*, 569–588.

Galli, M., & Guzzetti, F. (2007). Landslide vulnerability criteria: A case study from Umbria, Central Italy. *Environmental Management, 40*, 649–664.

Galli, M., Ardizzone, F., Cardinali, M., Guzzetti, F., & Reichenbach, P. (2008). Comparing landslide inventory maps. *Geomorphology, 94*, 268–289.

Glade, T. (1998). Establishing the frequency and magnitude of landslide-triggering rainstorm events in New Zealand. *Environmental Geology, 35*(2–3), 160–174.

Glade, T. (2000). Modelling landslide-triggering rainfalls in different regions of New Zealand – The soil water status model. *Zeitschrift für Geomorphologie NF, Suppl.-Bd. 122*, 63–84.

Glade, T. (2003). Vulnerability assessment in landslide risk analysis. *Die Erde, 134*(2), 123–146.

Glade, T., & Crozier, M. J. (2005). The nature of landslide hazard impact. In T. Glade, M. G. Anderson, & M. J. Crozier (Eds.), *Landslide hazard and risk* (pp. 43–74). Chichester: Wiley.

Godt, J. W. (1999). *Maps showing locations of damaging landslides caused by El Niño rainstorms, winter season 1997–98, San Francisco Bay region, California.* Pamphlet Field Studies Maps MF-2325-A-J (12 pp). Reston: U.S. Geological Survey.

Godt, J. W., Baum, R. L., & Chleborad, A. F. (2006). Rainfall characteristics for shallow landsliding in Seattle, Washington, USA. *Earth Surface Processes and Landforms, 31*, 97–110.

Godt, J. W., Baum, R. L., & Lu, N. (2009). Landsliding in partially saturated materials. *Geophysical Research Letters, 36*, L02403. doi:10.1029/2008GL035996.

Godt, J. W., Şener-Kaya, B., Lu, N., & Baum, R. L. (2012). Stability of infinite slopes under transient partially saturated seepage conditions. *Water Resources Research, 48*(5), W05505. doi: 10.1029/2011WR011408.

Gokceoglu, C., & Sezer, E. (2012). Soft computing modeling in landslide susceptibility assessment. In B. Pradhan, & M. Buchroithner (Eds.), *Terrigenous mass movements: Detection, modelling, early warning and mitigation using geoinformation technology* (pp. 51–90). Berlin: Springer.

Grossmann, J. (1998). Verfahren zur Berechnung der Grundwasserneubildung aus Niederschlag für große Einzugsgebiete. *gfw-Wasser/Abwasser, 139*, 14–23.

Guillard, C., & Zezere, J. (2012). Landslide susceptibility assessment and validation in the framework of municipal planning in Portugal: The case of Loures municipality. *Environmental Management, 50*, 721–735.

GuiYun, J., Yuan, T., Yu, L., & Yi, Z. (2008). A static and dynamic factors-coupled forecasting model of regional rainfall-induced landslides: A case study of Shenzhen. *Science in China Series E: Technological Sciences, 51*(Supplement 2), 164–175.

Guswa, A. J., Celia, M. A., & Rodriguez-Iturbe, I. (2002). Models of soil moisture dynamics in ecohydrology: A comparative study. *Water Resources Research, 38*(9), 5-1–5-15.

Guzzetti, F. (2005). *Landslide hazard and risk assessment* (389 pp). PhD Thesis, Rheinische Friedrich-Wilhelms-Universität Bonn, Mathematisch-Naturwissenschaftliche Fakultät. Retrieved July 13, 2014, from http://hss.ulb.uni-bonn.de/2006/0817/.

Guzzetti, F., Carrara, A., Cardinali, M., & Reichenbach, P. (1999). Landslide hazard evaluation: A review of current techniques and their application in a multi-scale study, Central Italy. *Geomorphology, 31*, 181–216.

Guzzetti, F., Cardinali, M., Reichenbach, P., & Carrara, A. (2000). Comparing landslide maps: A case study in the Upper Tiber River Basin, Central Italy. *Environmental Management, 25*(3), 247–263.

Guzzetti, F., Reichenbach, P., Cardinali, M., Galli, M., & Ardizzone, F. (2005). Probabilistic landslide hazard assessment at the basin scale. *Geomorphology, 72*, 272–299.

Guzzetti, F., Reichenbach, P., Ardizzone, F., Cardinali, M., & Galli, M. (2006). Estimating the quality of landslide susceptibility models. *Geomorphology, 81*, 166–184.

Guzzetti, F., Peruccacci, S., Rossi, M., & Stark, C. P. (2007). Rainfall thresholds for the initiation of landslides in central and southern Europe. *Meteorology and Atmospheric Physics, 98*, 239–267.

Guzzetti, F., Peruccacci, S., Rossi, M., & Stark, C. P. (2008). The rainfall intensity-duration control of shallow landslides and debris flows: An update. *Landslides, 5*, 3–17.

Guzzetti, F., Mondini, A. C., Cardinali, M., Fiorucci, F., Santangelo, M., & Chang, K.-T. (2012). Landslide inventory maps: New tools for an old problem. *Earth-Science Reviews, 112*, 42–66.

Hallegatte, S., & Przyluski, V. (2010). *The economics of natural disasters: Concepts and methods* (29 pp). Policy Research Working Paper 5507. Washington D.C.: The World Bank.

Hardenbicker, U., & Grunert, J. (2001). Temporal occurrence of mass movements in the Bonn area. *Zeitschrift für Geomorphologie NF, Suppl.-Bd. 125*, 13–24.

Haude, W. (1955). Zur Bestimmung der Verdunstung auf möglichst einfache Weise. *Mitteilungen des Deutschen Wetterdienstes, 11*(2), 24 S.

Hawke, R., & McConchie, J. (2011). In situ measurement of soil moisture and pore-water pressures in an 'incipient' landslide: Lake Tutira, New Zealand. *Journal of Environmental Management, 92*, 266–274.

He, Y., & Beighley, R. E. (2008). GIS-based regional landslide susceptibility mapping: A case study in southern California. *Earth Surface Processes and Landforms, 33*, 380–393.

Healy, R. W. (2010). *Estimating groundwater recharge*. Cambridge: Cambridge University Press. 256 pp.

Hearn, G. J., Hunt, T., Aubert, J., & Howell, J. H. (2008). *Landslide impacts on the road network of Lao PDR and the feasibility of implementing a slope management programme*. International Conference on Management of Landslide Hazard in the Asia-Pacific Region, Sendai, Japan.

Hearn, G. J., Hunt, T., & d'Agostino, S. (2011). Soil slope stabilization. In G. J. Hearn (Ed.), *Slope engineering for mountain roads* (pp. 165–188). Engineering Geology Special Publication 24. London: Geological Society.

Hergesell, M. (2003). *GIS-based modelling of regional groundwater recharge in Hesse, Germany* (102 S). Hydrologie in Hessen, Heft 1. Wiesbaden: Hessisches Landesamt für Umwelt und Geologie.

Hervás, J., & Bobrowsky, P. (2009). Mapping: Inventories, susceptibility, hazard and risk. In K. Sassa, & P. Canuti (Eds.), *Landslides – Disaster Risk Reduction* (pp. 321–349). Berlin: Springer.

Highland, L. M. (2006). *Estimating landslide losses – Preliminary results of a seven-state pilot project* (11 pp). Open-File Report 2006–1032. Reston: U.S. Geological Survey.

Highland, L. M. (2012). *Landslides in Colorado, USA – Impacts and loss estimation for 2010* (49 pp). Open-File Report 2012–1204. Reston: U.S. Geological Survey.

Hilker, N., Badoux, A., & Hegg, C. (2009). The Swiss flood and landslide damage database 1972–2007. *Natural Hazards and Earth System Sciences, 9*, 913–925.

HLfU. (1995). *Ermittlung der Grundwasserneubildung aus Niederschlag im Hessischen Ried*. Wiesbaden: Hessische Landesanstalt für Umwelt. 30 S.

Hölting, B. (1996). *Hydrologie: Einführung in die Allgemeine und Angewandte Hydrologie*. Stuttgart: Enke, 441 S.

Hong, Y., Adler, R., & Huffman, G. (2007). Use of satellite remote sensing data in the mapping of global landslide susceptibility. *Natural Hazards, 43*, 245–256.

Hutchinson, J. N. (1995). Keynote paper: Landslide hazard assessment. In D. H. Bell (Ed.), *Proceedings of the 6th International Symposium on Landslides* (Vol. 3, pp. 1805–1841). Rotterdam: Balkema.

Ibsen, M.-L., & Casagli, N. (2004). Rainfall patterns and related landslide incidence in the Porretta-Vergato region, Italy. *Landslides, 1*, 143–150.

Iverson, R. M. (2000). Landslide triggering by rain infiltration. *Water Resources Research, 36*(7), 1897–1910.

Jäger, D., Sandmeier, C., Schwindt, D., & Terhorst, B. (2013). Geomorphological and geophysical analyses in a landslide area near Ebermannstadt, Northern Bavaria. *E&G Quaternary Science Journal, 62*, 150–161.

Jaiswal, P., van Westen, C. J., & Jetten, V. (2010). Quantitative assessment of direct and indirect landslide risk along transportation lines in southern India. *Natural Hazards and Earth System Sciences, 10*, 1253–1267.

Jaiswal, P., van Westen, C. J., & Jetten, V. (2011). Quantitative estimation of landslide risk from rapid debris slides on natural slopes in the Nilgiri hills, India. *Natural Hazards and Earth System Sciences, 11*, 1723–1743.

Jakob, M., & Weatherly, H. (2003). A hydroclimatic threshold for landslide initiation on the North Shore Mountains of Vancouver, British Columbia. *Geomorphology, 54*, 137–156.

Jakob, M., Holm, K., Lange, O., & Schwab, J. W. (2006). Hydrometeorological thresholds for landslide initiation and forest operation shutdowns on the north coast of British Columbia. *Landslides, 3*, 228–238.

Jones, D. K. C. (1992). Landslide hazard assessment in the context of development. In G. J. H. McCall, D. J. C. Laming, & S. C. Scott (Eds.), *Geohazards: Natural and man-made* (pp. 117–141). London: Chapman & Hall.

Jordan, U. (1993). Die holozänen Massenverlagerungen des Wutachgebietes (Südschwarzwald). *Tübinger Geowissenschaftliche Arbeiten, C16,* 132 S.

Kanungo, D. P., Arora, M. K., Sarkar, S., & Gupta, R. P. (2006). A comparative study of conventional, ANN black box, fuzzy and combined neural and fuzzy weighting procedures for landslide susceptibility zonation in Darjeeling Himalayas. *Engineering Geology, 85*(3–4), 347–366.

Kanungo, D. P., Arora, M. K., Sarkar, S., & Gupta, R. P. (2009). Landslide susceptibility zonation (LSZ) mapping – A review. *Journal of South Asia Disaster Studies, 2*(1), 81–105.

Kirschbaum, D. B., Adler, R., Hong, Y., Hill, S., & Lerner-Lam, A. (2010). A global landslide catalog for hazard applications: Method, results, and limitations. *Natural Hazards, 52*, 561–575.

Kjekstad, O., & Highland, L. (2009). Economic and social impacts of landslides. In K. Sassa, & P. Canuti (Eds.), *Landslides – Disaster Risk Reduction* (pp. 573–587). Berlin: Springer.

Klimeš, J., & Blahůt, J. (2012). Landslide risk analysis and its application in regional planning: An example from the highlands of the Outer Western Carpathians, Czech Republic. *Natural Hazards, 64*, 1779–1803.

Klose, M., Damm, B., & Gerold, G. (2012a). Analysis of landslide activity and soil moisture in hillslope sediments using a landslide database and a soil water balance model. *GEO-ÖKO, 33*(3–4), 204–231.

Klose, M., Damm, B., Terhorst, B., Schulz, N., & Gerold, G. (2012b). Wirtschaftliche Schäden durch gravitative Massenbewegungen. Entwicklung eines empirischen Berechnungsmodells mit regionaler Anwendung. *Interpraevent, 12*, 979–990.

Klose, M., Gruber, D., Damm, B., & Gerold, G. (2014a). Spatial databases and GIS as tools for regional landslide susceptibility modeling. *Zeitschrift für Geomorphologie NF, 58*(1), 1–36.

Klose, M., Damm, B., & Terhorst, B. (2014b). Landslide cost modeling for transportation infrastructures: A methodological approach. *Landslides*. doi:10.1007/s10346-014-0481-1.

Klose, M., Highland, L., Damm, B., & Terhorst, B. (2014c). Estimation of direct landslide costs in industrialized countries: Challenges, concepts, and case study. In K. Sassa, P. Canuti, & Y. Yin (Eds.), *Landslide Science for a Safer Geoenvironment, Vol. 2: Methods of Landslide Studies* (pp. 661–667). Berlin: Springer.

Knoblich, K. (1967). Mechanische Gesetzmäßigkeiten beim Auftreten von Hangrutschungen. *Zeitschrift für Geomorphologie NF, 11*, 286–299.

Ko Ko, C., Chowdhury, R., & Flentje, P. (2005). Hazard and risk assessment of rainfall-induced landsliding along a railway line. *Quarterly Journal of Engineering Geology and Hydrogeology, 38*, 197–213.

Kobashi, S., & Suzuki, M. (1988). Hazard index for the judgment of slope stability in the Rokko Mountain region. In *Proceedings Interpraevent 1988, Volume 1* (pp. 223–233).

Kraus, H. (2004). *Die Atmosphäre der Erde: Eine Einführung in die Meteorologie*. Berlin: Springer. 422 S.

Krauter, E. (1992). Hangrutschungen – Ein Umweltproblem. In H. J. Matthias, A. Grün (Hrsg.), *Ingenieurvermessung 92. Beiträge zum XI. Internationalen Kurs für Ingenieurvermessung 2* (pp. V4/1–V4/12). Bonn: Ferd. Dümmlers.

Krauter, E., & Steingötter, K. (1983). Die Hangstabilitätskarte des linksrheinischen Mainzer Beckens. *Geologisches Jahrbuch C, 34*, 3–31.

Kreibich, H., van den Bergh, J. C. J. M., Bouwer, L. M., Bubeck, P., Ciavola, P., Green, C., et al. (2014). Costing natural hazards. *Nature Climate Change, 4*, 303–306.

Krohn, J. P., & Slosson, J. E. (1976). Landslide potential in the United States. *California Geology, 29*(10), 224–231.

Lascano, R. J. (1991). Review of models for predicting soil water balance. In M. V. K. Sivakumar, J. S. Wallace, C. Renard, & C. Giroux (Eds.), *Soil water balance in the Sudano-Sahelian Zone* (pp. 443–458). IAHS Publication No. 199.

Lateltin, O., Haemmig, C., Raetzo, H., & Bonnard, C. (2005). Landslide risk management in Switzerland. *Landslides, 2*, 313–320.

Lee, E. M., & Jones, D. K. C. (2004). *Landslide risk assessment*. London: Thomas Telford. 464 pp.

Lee, S., & Min, K. (2001). Statistical analysis of landslide susceptibility at Yongin, Korea. *Environmental Geology, 40*, 1095–1113.

Lee, S., Choi, J., & Min, K. (2002). Landslide susceptibility analysis and verification using the Bayesian probability model. *Environmental Geology, 43*, 120–131.

Li, Z. H., Huang, H. W., Xue, Y. D., & Yin, J. (2009). Risk assessment of rockfall hazards on highways. *Georisk, 3*(3), 147–154.

Link, M. (1998). *Reliefentwicklung und Oberflächenformung im Linzgau unter besonderer Berücksichtigung von Massenverlagerungen* (236 S). Dissertation, Albert-Ludwigs-Universität Freiburg.

Lu, N., & Godt, J. W. (2013). *Hillslope hydrology and stability*. Cambridge: Cambridge University Press. 458 pp.

Lu, N., & Likos, W. J. (2004). *Unsaturated soil mechanics*. Hoboken: Wiley. 556 pp.

MacLeod, A., Hofmeister, R. J., Wang, Y., & Burns, S. (2005). *Landslide indirect losses: Methods and case studies from Oregon* (14 pp). Open-File Report O-05-X. Portland: Oregon Department of Geology and Mineral Industries.

Magliulo, P., Di Lisio, A., Russo, F., & Zelano, A. (2008). Geomorphology and landslide susceptibility assessment using GIS and bivariate statistics: A case study in southern Italy. *Natural Hazards, 47*, 411–435.

Mancini, F., Ceppi, C., & Ritrovato, G. (2010). GIS and statistical analysis for landslide susceptibility mapping in the Daunia area, Italy. *Natural Hazards and Earth System Sciences, 10*, 1851–1864.

Mathew, J., Jha, V. K., & Rawat, G. S. (2009). Landslide susceptibility zonation mapping and its validation in part of Garhwal Lesser Himalaya, India, using binary logistic regression analysis and receiver operating characteristic curve method. *Landslides, 6*, 17–26.

Mathur, H. N. (1982). Influence of human activities on landslides, mudflows and slope movements in India and efforts at reducing their negative impact. In A. Sheko (Ed.), *Report of the International Seminar on Landslides and Mudflows and the Prevention of Their Negative Impact on the Environment* (pp. 20–44). October 1981, Alma-Ata (USSR). Moscow: Centre of International Projects, GKNT.

Meyer, V., Becker, N., Markantonis, V., Schwarze, R., van den Bergh, J. C. J. M., Bouwer, L. M., et al. (2013). Review article: Assessing the costs of natural hazards – State of the art and knowledge gaps. *Natural Hazards and Earth System Sciences, 13*, 1351–1373.

Michoud, C., Derron, M.-H., Horton, P., Jaboyedoff, M., Baillifard, F.-J., Loye, A., et al. (2012). Rockfall hazard and risk assessments along roads at a regional scale: Example in Swiss Alps. *Natural Hazards and Earth System Sciences, 12*, 615–629.

Monteith, J. L. (1976). *Vegetation and the Atmosphere*. Vol. 2, Case Studies. London: Academic Press. 459 pp.

Mousavi, S. M., Omidvar, B., Ghazban, F., & Feyzi, R. (2011). Quantitative risk analysis for earthquake-induced landslides – Emamzadeh Ali, Iran. *Engineering Geology, 122*, 191–203.

Nadim, F., Kjekstad, O., Peduzzi, P., Herold, C., & Jaedicke, C. (2006). Global landslide and avalanche hotspots. *Landslides, 3*, 159–173.

Nandi, A., & Shakoor, A. (2009). A GIS-based landslide susceptibility evaluation using bivariate and multivariate statistical analyses. *Engineering Geology, 110*, 11–20.

Negi, I. S., Kumar, K., Kathait, A., & Prasad, P. S. (2013). Cost assessment of losses due to recent reactivation of Kaliasaur landslide on National Highway 58 in Garwhal Himalaya. *Natural Hazards, 68*, 901–914.

Neuhäuser, B., & Terhorst, B. (2007). Landslide susceptibility assessment using "weights-of-evidence" applied to a study area at the Jurassic escarpment (SW-Germany). *Geomorphology, 86*, 12–24.

Neuhäuser, B., Damm, B., & Terhorst, B. (2012a). GIS-based assessment of landslide susceptibility on the base of the weights-of-evidence model. *Landslides, 9*, 511–528.

Neuhäuser, B., Terhorst, B., & Damm, B. (2012b). Landslide identification and modelling in flysch areas of the European Alpine Foreland. *Zeitschrift für Geomorphologie NF, 56*(4), 115–146.

Neuhäuser, B., Damm, B., & Terhorst, B. (2013). Modellierung der Hangrutschungsgefährdung unter dem Aspekt von Klimaänderungen am Beispiel Nördlicher Wiener wald. *Berichte Geol. B.-A., 100*, 61–65.

Oh, H.-J., Park, N.-W., Lee, S.-S., & Lee, S. (2012). Extraction of landslide-related factors from ASTER imagery and its application to landslide susceptibility mapping. *International Journal of Remote Sensing, 33*, 3211–3231.

Ohara, J., Fujisawa, K., Ikeda, M., Ueno, Y., Kokuryou, N., & Kasahara, R. (2008). Case study of estimation of financial loss by landslide disaster. In *Proceedings of the First World Landslide Forum, Poster Session Volume* (pp. 73–75). Tokyo, Japan, 18–21 November 2008.

Ohlmacher, G. C., & Davis, J. D. (2003). Using multiple logistic regression and GIS technology to predict landslide hazard in northeast Kansas, USA. *Engineering Geology, 69*, 331–343.

Ojeda-Moncayo, J., Locat, J., Couture, R., & Leroueil, S. (2004). The magnitude of landslides: An overview. In W. Lacerda, M. Ehrlich, S. A. B. Fontoura, & A. S. F. Sayao (Eds.), *Landslides: Evaluation and stabilization* (Vol. 1, pp. 379–384). London: Taylor & Francis Group.

Papathoma-Köhle, M., Kappes, M., Keiler, M., & Glade, T. (2011). Physical vulnerability assessment for alpine hazards: State of the art and future needs. *Natural Hazards, 58*, 645–680.

Pardeshi, S. D., Autade, S. E., & Pardeshi, S. S. (2013). Landslide hazard assessment: Recent trends and techniques. *SpringerPlus, 2013*(2), 523. doi:10.1186/2193-1801-2-523.

Parise, M. (2001). Landslide mapping techniques and their use in the assessment of the landslide hazard. *Physics and Chemistry of the Earth C, 26*(9), 697–703.

Pasuto, A., & Silvano, S. (1998). Rainfall as a trigger of shallow mass movements. A case study in the Dolomites, Italy. *Environmental Geology, 35*, 184–189.

Penman, H. L. (1948). Natural evaporation from open water, bare soil and grass. *Proceedings of the Royal Meteorological Society, 193*, 120–145.

Pereira, S., Zêzere, J. L., & Bateira, C. (2012). Technical note: Assessing predictive capacity and conditional independence of landslide predisposing factors for shallow landslide susceptibility models.*Natural Hazards and Earth System Sciences, 12*, 979–988.

Petley, D. N. (2012). Global patterns of loss of life from landslides. *Geology, 40*, 927–930.

Picarelli, L., Oboni, F., Evans, S. G., Mostyn, G., & Fell, R. (2005). Hazard characterization and quantification. In O. Hungr, R. Fell, R. Couture, & E. Eberhardt (Eds.), *Landslide risk management* (pp. 27–61). London: Taylor & Francis Group.

Pimenta, M. T. (2000). Water balances using GIS. *Physics and Chemistry of the Earth B, 25*(7–8), 695–698.

Pitilakis, K., Fotopoulou, S., Argyroudis, S., Pitilakis, D., Senetakis, K., Treulopoulos, K., Kakderi, K., & Riga, E. (2011). *Physical vulnerability of elements at risk to landslides: Methodology for evaluation, fragility curves and damage states for buildings and lifelines* (195 pp). SafeLand FP7, Deliverable D2.5. Retrieved August 8, 2014, from http://www.safeland-fp7.eu/results/Documents/D2.5_revised.pdf.

Ponziani, F., Pandolfo, C., Stelluti, M., Berni, N., Brocca, L., & Moramarco, T. (2012). Assessment of rainfall thresholds and soil moisture modeling for operational hydrogeological risk prevention in the Umbria region (central Italy). *Landslides, 9*, 229–237.

Popescu, M. E. (1994). A suggested method for reporting landslide causes. *Bulletin of the International Association of Engineering Geology, 50*(1), 71–74.

Prokešová, R., Medveďová, A., Tábořík, P., & Snipková, Z. (2013). Towards hydrological triggering mechanisms of large deep-seated landslides. Landslides, *10*, 239–254.

Public Works Department Malaysia. (2008). National slope master plan 2009–2023, Appendix B – Landslide inventory and total estimated costs. Retrieved November 13, 2013, from http://slopes.jkr.gov.my/Documentation/NSMP/English%20Version/NSMPSecApendixB.pdf.

Raetzo, H., Lateltin, O., Bollinger, D., & Tripet, J. P. (2002). Hazard assessment in Switzerland – Codes of practice for mass movements. *Bulletin of Engineering Geology and the Environment, 61*(3), 263–268.

Rahman, A., Khan, A. N., Collins, A. E., & Qazi, F. (2011). Causes and extent of environmental impacts of landslide hazard in the Himalayan region: A case study of Murree, Pakistan. *Natural Hazards, 57*, 413–434.

Ranatunga, K., Nation, E. R., & Barrat, D. G. (2008). Review of soil water models and their applications in Australia. *Environmental Modelling and Software, 23*, 1182–1206.

Ray, R. L., & Jacobs, J. M. (2007). Relationships among remotely sensed soil moisture, precipitation and landslide events. *Natural Hazards, 43*, 211–222.

Ray, R. L., Jacobs, J. M., & Cosh, M. H. (2010). Landslide susceptibility mapping using downscaled AMSR-E soil moisture: A case study from Cleveland Corral, California, US. *Remote Sensing of Environment, 114*, 2624–2636.

Regmi, N. R., Giardino, J. R., & Vitek, J. D. (2010). Modeling susceptibility to landslides using the weight of evidence approach: Western Colorado, USA. *Geomorphology, 115*, 172–187.

Reid, L. M., & Page, M. J. (2002). Magnitude and frequency of landsliding in a large New Zealand catchment. *Geomorphology, 49*, 71–88.

Reid, M. E., Baum, R. L., LaHusen, R. G., & Ellis, W. L. (2008). Capturing landslide dynamics and hydrologic triggers using near-real-time monitoring. In Z. Chen, J. Zhang, Z. Li, F. Wu, & K. Ho (Eds.), *Landslides and Engineered slopes: From the Past to the Future* (Vol. 1, pp. 179–191). London: Taylor & Francis Group.

Remondo, J., González, A., Díaz De Terán, J. R., Cendrero, A., Fabbri, A., & Chung, C.-J. F. (2003). Validation of landslide susceptibility maps; examples and applications from a case study in Northern Spain. *Natural Hazards, 30*, 437–449.

Remondo, J., Bonachea, J., & Cendrero, A. (2005). A statistical approach to landslide risk modelling at basin scale: From landslide susceptibility to quantitative risk assessment. *Landslides, 2*, 321–328.

Remondo, J., Bonachea, J., & Cendrero, A. (2008). Quantitative landslide risk assessment and mapping on the basis of recent occurrences. *Geomorphology, 94*, 496–507.

Saito, H., Nakayama, D., & Matsuyama, H. (2010). Relationship between the initiation of a shallow landslide and rainfall intensity–duration thresholds in Japan. *Geomorphology, 118*, 167–175.

Salciarini, D., Godt, J. W., Savage, W. Z., Baum, R. L., & Conversini, P. (2008). Modeling landslide recurrence in Seattle, Washington, USA. *Engineering Geology, 102*, 227–237.

Santacana, N., Baeza, B., Corominas, J., De Paz, A., & Marturiá, J. (2003). A GIS-based multivariate statistical analysis for shallow landslide susceptibility mapping in La Pobla de Lillet Area (Eastern Pyrenees, Spain). *Natural Hazards, 30*, 281–295.

Scalon, B. R., Healy, R. W., & Cook, P. G. (2002). Choosing appropriate techniques for quantifying groundwater recharge. *Hydrogeology Journal, 10*, 18–39.

Schmidt, J., & Dikau, R. (2004). Modeling historical climate variability and slope stability. *Geomorphology, 60*, 433–447.

Schunke, E. (1971). Die Massenverlagerungen an den Schichtstufen und Schichtkämmen des Weser-Leine-Berglandes. *Nachrichten der Akademie der Wissenschaften in Göttingen. II. Mathematisch-Physikalische Klasse, 3*, 1–35.

Schuster, R. L. (1978). Introduction. In R. L. Schuster, & R. J. Krizek (Eds.), *Landslides: Analysis and control* (pp. 1–10). Transportation Research Board, Special Report 176. Washington D.C.: National Academy of Sciences.

Schuster, R. L. (1996). Socioeconomic significance of landslides. In A. K. Turner, & R. L. Schuster (Eds.), *Landslides: Investigation and mitigation* (pp. 12–35). Transportation Research Board, Special Report 247. Washington D.C.: National Academy Press.

Schuster, R. L., & Fleming, R. W. (1986). Economic losses and fatalities due to landslides. *Bulletin of the Association of Engineering Geologists, 23*(1), 11–28.

Schuster, R. L., & Highland, L. M. (2001). *Socioeconomic and environmental impacts of landslides in the western hemisphere*. Open-File Report 01-0276, U.S. Geological Survey. Retrieved July 15, 2014, from http://pubs.usgs.gov/of/2001/ofr-01-0276/.

Schuster, R. L., & Highland, L. M. (2007). The third Hans Cloos lecture. Urban landslides: Socioeconomic impact and overview of mitigative strategies. *Bulletin of Engineering Geology and the Environment, 66*, 1–27.

Schwab, J. C., Gori, P. L., & Jeer, S. (Eds.) (2005). *Landslide hazards and planning*. PAS Report 533/534. Washington D.C.: American Planning Association. 208 pp.

Shahabi, H., Ahmad, B. B., & Khezri, S. (2013). Evaluation and comparison of bivariate and multivariate statistical methods for landslide susceptibility mapping (case study: Zab basin). *Arabian Journal of Geosciences, 6*(10), 3885–3907.

Shearer, C. F., Taylor, F. A., & Fleming, R. W. (1983). *Distribution and costs of landslides in San Diego County, California, during the rainfall years of 1978–79 and 1979–80* (15 pp). Open-File Report 83-582. Reston: U.S. Geological Survey.

Sidle, R. C., & Ochiai, H. (2006). *Landslides: Processes, prediction, and land use*. Water Resources Monograph 18. Washington D.C.: American Geophysical Union. 312 pp.

Silva, A., & Pereira, S. (2014). Assessment of physical vulnerability and potential losses of buildings due to shallow slides. *Natural Hazards, 72*, 1029–1050.

Smith, R. (1958). Economic and legal aspects. In E. B. Eckel (Ed.), *Landslides and engineering practice* (pp. 6–19). Highway Research Board, Special Report 28. Washington D.C.: National Academy of Sciences.

Soeters, R., & Van Westen, C. J. (1996). Slope instability recognition, analysis and zonation. In A. K. Turner, & R. L. Schuster (Eds.), *Landslides: Investigation and mitigation* (pp. 129–177). Transportation Research Board, Special Report 247. Washington D.C.: National Academy Press.

Sophocleous, M. A. (1991). Combining the soilwater balance and water-level fluctuation methods to estimate natural groundwater recharge: Practical aspects. *Journal of Hydrology, 124*, 229–241.

Sterlacchini, S., Frigerio, S., Giacomelli, P., & Brambilla, M. (2007). Landslide risk analysis: A multi-disciplinary methodological approach. *Natural Hazards and Earth System Sciences, 7*, 657–675.

Süzen, M. L., & Doyuran, V. (2004a). Data driven bivariate landslide susceptibility assessment using geographical information systems: A method and application to Asarsuyu catchment, Turkey. *Engineering Geology, 71*, 303–321.

Süzen, M. L., & Doyuran, V. (2004b). A comparison of the GIS based landslide susceptibility assessment methods: Multivariate versus bivariate. *Environmental Geology, 45*, 665–679.

Szabó, J. (2003). The relationship between landslide activity and weather: Examples from Hungary. *Natural Hazards and Earth System Sciences, 3*, 43–52.

Taylor, F. A., & Brabb, E. E. (1972). *Maps showing distribution and cost by counties of structurally damaging landslides in the San Francisco Bay region, California, winter of 1968–69*. USGS Miscellaneous Field Studies Map: 327.

Terhorst, B., Paeth, H., Baumhauer, R., & Schwindt, D. (2013). Massenbewegungen: Die unterschätzte Naturgefahr in Mittelgebirgsregionen – Aktuelle Rutschereignisse und Wetterlage im Frühjahr 2013. In *Tagungsband der Jahrestagung des AK Geomorphologie* (S. 35). Eichstätt, 7.-9. Oktober 2013.

Terlien, M. T. J. (1998). The determination of statistical and deterministic hydrological landslide-triggering thresholds. *Environmental Geology, 35*(2–3), 124–130.

Theisen, H.-W. (1998). Der Einfluß der Bodenfeuchte auf die Sicherheit rutschungsgefährdeter Hänge in schwach bindigen Böden. *Mitteilungen des Instituts für Hydrologie und Wasserwirtschaft der Universität Karlsruhe, 62,* 190 S.

Thiery, Y., Malet, J.-P., Sterlacchini, S., Puissant, A., & Maquaire, O. (2007). Landslide susceptibility assessment by bivariate methods at large scales: Application to a complex mountainous environment. *Geomorphology, 92*, 38–59.

Trezzini, F., Giannella, G., & Guida, T. (2013). Landslide and flood: Economic and social impact in Italy. In C. Margottini, P. Canuti, & K. Sassa (Eds.), *Landslide Science and Practice, Vol. 7: Social and Economic Impact and Policies* (pp. 171–176). Berlin: Springer.

Tsai, T. L., & Chen, H. F. (2010). Effects of degree of saturation on shallow landslides triggered by rainfall. *Environmental Earth Sciences, 59*(6), 1285–1295.

Turner, A. K., & McGuffey, V. C. (1996). Organization of investigation process. In A. K. Turner, & R. L. Schuster (Eds.), *Landslides: Investigation and mitigation* (pp. 121–128). Transportation Research Board, Special Report 247. Washington D.C.: National Academy Press.

USGS. (2013). Landslide inventory pilot project – Kentucky. U.S. Geological Survey. Retrieved November 20, 2013, from http://landslides.usgs.gov/regional/inventory/kentucky/.

Uzielli, M., Nadim, F., Lacasse, S., & Kaynia, A. M. (2008). A conceptual framework for quantitative estimation of physical vulnerability to landslides. *Engineering Geology, 102*, 251–256.

Van Asch, Th. W. J., Buma, J., & Van Beek, L. P. H. (1999). A view on some hydrological triggering systems in landslides. *Geomorphology, 30*, 25–32.

Van Beek, L. P. H., & Van Asch, Th. W. J. (2004). Regional assessment of the effects of land-use change on landslide hazard by means of physically based modelling. *Natural Hazards, 31*, 289–304.

Van Beurden, S. (1997). Hydrology, soil mechanics and kinematics of slow mass movements in the Widenbach catchment, Switzerland. *Nederlandse Geografische Studies, 225*, Univ. Utrecht, 253 pp.

Van Den Eeckhaut, M., Vanwalleghem, T., Poesen, J., Govers, G., Verstraeten, G., & Vandekerckhove, L. (2006). Prediction of landslide susceptibility using rare events logistic regression: A case-study in the Flemish Ardennes (Belgium). *Geomorphology, 76*, 392–410.

Van Den Eeckhaut, M., Hervás, J., & Montanarella, L. (2010). Validation of landslide inventory, susceptibility and hazard maps for risk management. In C. Corbane, D. Carrion, M. Broglia, & M. Pesaresi (Eds.), *Proceedings 2nd International Workshop on Validation of Geo-Information Products for Crisis Management* (pp. 41–49). EUR 24530 EN. Luxembourg: Publications Office of the European Union.

Van Westen, C. J. (1993). *Application of geographic information systems to landslide hazard zonation* (245 pp). ITC Publication 15, International Institute for Aerospace Survey and Earth Sciences (ITC). Enschede, The Netherlands.

Van Westen, C. J. (2000). The modelling of landslide hazards using GIS. *Surveys in Geophysics, 21*, 241–255.

Van Westen, C. J. (2004). Geo-information tools for landslide risk assessment. An overview of recent developments. In W. Lacerda, M. Ehrlich, S. A. B Fontoura, & A. S. F. Sayao (Eds.), *Landslides: Evaluation and stabilization* (Vol. 1, pp. 39–56). London: Taylor & Francis Group.

Van Westen, C. J., Van Asch, T. W. J., & Soeters, R. (2006). Landslide hazard and risk zonation – Why is it still so difficult? *Bulletin of Engineering Geology and the Environment, 65*, 167–184.

Van Westen, C. J., Castellanos, E., & Kuriakose, S. L. (2008). Spatial data for landslide susceptibility, hazard, and vulnerability assessment: An overview. *Engineering Geology, 102*, 112–131.

Varnes, D. J., & IAEG (1984). *Landslide hazard zonation: A review of principles and practice* Paris: UNESCO. 63 pp.

Vranken, L., Van Turnhout, P., Van Den Eeckhaut, M., Vandekerckhove, L., & Poesen, J. (2013). Economic valuation of landslide damage in hilly regions: A case study from Flanders, Belgium. *Science of the Total Environment, 447*, 323–336.

Walkinshaw, J. (1992). Landslide correction costs on US state highway systems. *Transportation Research Record, 1343*, 36–41.

Wang, H. B., & Sassa, K. (2005). Comparative evaluation of landslide susceptibility in Minamata area, Japan. *Environmental Geology, 47*, 956–966.

Wang, Y., Summers, R. D., & Hofmeister, R. J. (2002). *Landslide loss estimation pilot project in Oregon* (47 pp). Open-File Report O-02-05. Portland: Oregon Department of Geology and Mineral Industries.

Wang, H., Liu, G., Xu, W., & Wang, G. (2005). GIS-based landslide hazard assessment: An overview. *Progress in Physical Geography, 29*, 548–567.

Wang, B., Ruel, M., Couture, R., VanDine, D., Bobrowsky, P., & Blais-Stevens, A. (2013). Current status of landslide guidelines around the world. In C. Margottini, P. Canuti, & K. Sassa (Eds.), *Landslide Science and Practice, Vol. 7: Social and Economic Impact and Policies* (pp. 271–277). Berlin: Springer.

Wolterstorff, E. (2002). *Ansätze einer monetären Bewertung volkswirtschaftlicher Schäden durch Rutschungen – Diskutiert an Beispielen aus Rheinland-Pfalz* (119 S. + Anhang). Diplomarbeit, Johannes Gutenberg-Universität Mainz, Geographisches Institut.

Wright, R. H., Campbell, R. H., & Nilsen, T. H. (1974). Preparation and use of isopleth maps of landslide deposits. *Geology, 2*, 483–485.

Wu, S., Shi, L., Wang, R., Tan, C., Hu, D., Mei, Y., & Xu, R. (2001). Zonation of the landslide hazards in the forereservoir region of the Three Gorges Project on the Yangtze River. *Engineering Geology, 59*, 51–58.

Wyoming Homeland Security. (2011). Wyoming multi-hazard mitigation plan – Landslides. Retrieved November 20, 2013, from http://hls.wyogov/Library/mit_plan/11_Landslide.pdf.

Xu, C. Y., & Singh, V. P. (1998). A review of monthly water balance models for water resources investigations. *Water Resources Management, 12*, 31–50.

Yalcin, A. (2008). GIS-based landslide susceptibility mapping using analytical hierarchy process and bivariate statistics in Ardesen (Turkey). *Catena, 72*, 1–12.

Yeh, H.-F., & Lee, C.-H. (2013). Soil water balance model for precipitation-induced shallow landslides. *Environmental Earth Sciences, 70*, 2691–2701.

Yin, K. L., & Yan, T. Z. (1988). Statistical prediction models for slope instability of metamorphosed rocks. In C. Bonnard (Ed.), *Proceedings of the 5th International Symposium on Landslides* (Vol. 2, pp. 1269–1272). Rotterdam: Balkema.

Zêzere, J. L. (2002). Landslide susceptibility assessment considering landslide typology. A case study in the area north of Lisbon (Portugal). *Natural Hazards and Earth System Sciences, 2*, 73–82.

Zêzere, J. L., Reis, E., Garcia, R., Oliveira, S., Rodrigues, M. L., Vieira, G., & Ferreira, A. B. (2004). Integration of spatial and temporal data for the definition of different landslide hazard scenarios in the area north of Lisbon (Portugal). *Natural Hazards and Earth System Sciences, 4*, 133–166.

Zêzere, J. L., Trigo, R. M., & Trigo, I. F. (2005). Shallow and deep landslides induced by rainfall in the Lisbon region (Portugal): Assessment of relationships with the North Atlantic Oscillation. *Natural Hazards and Earth System Sciences, 5*, 331–344.

Zêzere, J. L., Oliveira, S. C., Garcia, R. A. C., & Reis, E. (2007). Landslide risk analysis in the area North of Lisbon (Portugal): Evaluation of direct and indirect costs resulting from a motorway disruption by slope movements. *Landslides, 4*, 123–136.

Zêzere, J. L., Garcia, R. A. C., Oliveira, S. C., & Reis, E. (2008). Probabilistic landslide risk analysis considering direct costs in the area north of Lisbon (Portugal). *Geomorphology, 94*, 467–495.

Zhang, L., Walker, G. R., & Dawes, W. R. (2002). Water balance modelling: Concepts and applications. In T. R. McVicar, L. Rui, J. Walker, R. W. Fritzpatrick, & L. Changming (Eds.), *Regional water and soil assessment for managing sustainable agriculture in China and Australia* (pp. 31–47). ACIAR Monograph No. 84. Canberra: Australian Centre for International Agricultural Research.

Chapter 5
Results

5.1 Spatial Hazard—Where Do Landslides Occur?

Note: The entire section has been published in Klose et al. 2014a.

5.1.1 Landslide Characteristics and Geofactor Weights

5.1.1.1 Slope Gradient

Landslides primarily occur at slope gradients between 0° and 49°, showing a frequency maximum of 31 % in the class 21°–28°. A significant landslide predisposition, however, is only identified for slope gradients in the range of 21°–49°, as landslide densities of >0.80 and information values of >4.00 illustrate. Hillsides with slope gradient <14° are not susceptible to landslides and can be regarded as widely stable. This holds also for rock slopes steeper than 49°, which are not classified as landslide-prone, since they are not represented in the inventory (cf. Table 5.1).

5.1.1.2 Slope Curvature

Slopes with a high degree of plan curvature show a substantial tendency to mass movements. The information values of both strong concavity and convexity, which account for 2.65 and 3.22, attest these attributes the second highest importance on landslide predisposition. Nevertheless, the inventory data verifies that most slope failures take place on hillsides featuring low or even no curvature. The widespread presence of such slopes, however, causes low landslide densities, wherefore statistics discovers a rather stabilizing influence, as information values between −1.00 and 0.00 indicate (cf. Table 5.1).

M. Klose, *Landslide Databases as Tools for Integrated Assessment of Landslide Risk*, Springer Theses, DOI 10.1007/978-3-319-20403-1_5

Table 5.1 Landslide density measures and information values derived for the different attribute classes

Geofactors and attribute classes $A(i)$		Spatial extension* $S_{A(i)}$ (%)	Number of landslides $N_{A(i)}$	Number of landslides* $N_{A(i)}$ (%)	Landslide density (km^2) $N_{A(i)}/S_{A(i)}$	Information value $I_{A(i)}$
Geomorphometry						
Slope gradient	0°–7°	87.41	119	15.74	0.00	−1.71
	7°–14°	9.91	76	10.05	0.01	0.01
	14°–21°	1.96	168	22.22	0.16	2.43
	21°–28°	0.54	237	31.35	0.84	4.07
	28°–35°	0.14	117	15.48	1.57	4.69
	35°–42°	0.03	33	4.37	1.82	4.85
	42°–49°	0.01	6	0.79	1.61	4.72
	49°–90°	0.00	0	0.00	0.00	0.00
Slope curvature	Strong convex	0.21	39	5.16	0.36	3.22
	Convex	37.71	299	39.55	0.02	0.05
	No curvature	22.93	64	8.47	0.01	−1.00
	Concave	38.98	335	44.31	0.02	0.13
	Strong concave	0.18	19	2.51	0.20	2.65
Slope aspect	N	12.29	110	14.55	0.02	0.17
	NO	11.61	115	15.21	0.02	0.27
	O	12.28	124	16.40	0.02	0.29
	SO	10.93	75	9.92	0.01	−0.10
	S	11.80	81	10.71	0.01	−0.10
	SW	11.42	83	10.98	0.01	−0.04
	W	12.27	91	12.04	0.01	−0.02
	NW	11.32	77	10.19	0.01	−0.11
	No aspect	6.09	0	0.00	0.00	0.00
Lithology						
Quaternary lowland deposits		75.37	44	5.82	0.00	−2.56
Marlstone		4.88	62	8.20	0.02	0.52
Clay-/limestone		2.11	235	31.08	0.21	2.69
Sand-/limestone		14.68	411	54.37	0.05	1.31
Greywacke/clay shale		1.45	4	0.53	0.01	−1.01
Granite		0.17	0	0.00	0.00	0.00
Land use						
Forest		25.52	551	72.88	0.04	1.05
Grassland		22.55	66	8.73	0.01	−0.95

(continued)

Table 5.1 (continued)

Geofactors and attribute classes $A(i)$	Spatial extension* $S_{A(i)}$ (%)	Number of landslides $N_{A(i)}$	Number of landslides* $N_{A(i)}$ (%)	Landslide density (km^2) $N_{A(i)}/S_{A(i)}$	Information value $I_{A(i)}$
Arable land	38.20	47	6.22	0.00	−1.82
Urban area	8.15	92	12.17	0.02	0.40
Other land use type	5.58	0	0.00	0.00	0.00

*Based on the relevant geofactor
The table also highlights the spatial extension of each attribute class and its number of landslides (*Source* modified after Klose et al. 2014a)

5.1.1.3 Slope Aspect

Mass movements are distributed over the entire spectrum of the main directories. A slight dominance of slope instability is for the NE sector observable. South- and westward orientated hillsides, in contrast, show somewhat lower occurrence frequency. Slope aspect is generally no important factor in controlling landslide susceptibility. Information values in the range of 0.29 to −0.11 signalize this weak influence. According to the landslide distribution, positive information values are obtained for NE facing slopes, while negative ones are typical for slopes with SW exposition (cf. Table 5.1).

5.1.1.4 Lithology

Most landslides are related to Mesozoic sedimentary rock, especially to the attribute classes sand-/limestone and clay-/limestone, which include 235 and 411 mass movements, respectively. This prominence, however, is not clarified by the landslide density that possesses only intermediate values. Nevertheless, both attributes have a significant positive impact on landslide occurrence, as information values of 1.31 and 2.69 clearly document. On the other hand, the lithologic categories greywacke/clay shale and Quaternary lowland deposits are ascertained to be negatively associated with slope instability. The predicted stabilization effect is expressed by information values of −1.01 and −2.56, which must be discussed critically (cf. Sect. 5.1.4).

5.1.1.5 Land Use

A majority of 73 % of the recorded landslides is situated in areas that are in forestal use. Due to the large spatial extension of this land use type, its landslide density is with 0.04 comparatively low. This manifests in a slightly positive information value of 1.05, which proves a doubtful susceptibility of forest area to mass movements. Distinct lower landslide frequency is common for the other land use categories that have almost negligible density values. While a slightly positive landslide predisposition is identified for urban area, grassland and arable land are not defined to be landslide-prone, as information values of −0.95 and −1.82 indicate. These statistical weights, however, require a critical reflection (cf. Sect. 5.1.4).

5.1.2 Landslide Susceptibility Map

5.1.2.1 Categories of Landslide Susceptibility

The value of *SI*(*x*) ranges from −7.09 to 12.10 and is displayed in the final map in four categories, namely no, low, moderate and high predisposition. One important result of the modeling with this limited input data is that slope gradient constitutes a dominant controlling factor of mass movements (cf. Table 5.1). Therefore, the classification of the susceptibility scale is oriented towards the information values calculated for this geofactor. After taking into account other factor influences, there is clear evidence that a susceptibility index <3.00 still indicates stable conditions. For instance, even if slope gradient is between 0° and 7°, the total susceptibility of forest area with clay-/limestone bedrock is still >2.00, although slope instability is rather unlikely under these circumstances (cf. Table 5.1). To avoid that too much area is being classified as landslide-prone, the lower boundary of susceptibility is set to a value of 3.00. Index values above this threshold suggest a significant landslide predisposition, whose level is specified using three different classes with equal intervals. This guarantees coherent class occupancy and finally results in a less conservative delineation of susceptibility zones.

5.1.2.2 General Overview and Key Areas of Landslide Susceptibility

Landslide susceptibility in Lower Saxony is widely determined by the spatial pattern of the regional relief configuration (cf. Fig. 5.1). A major part of the study area, especially the lowland north of Hannover, is almost free of landslide-prone terrain. Large territory of the Lower Saxon Uplands, in contrast, shows significant landslide susceptibility. Three main distribution areas, which differ in their level and areal composition of susceptibility, can be divided. A clearly definable zone of moderate to high landslide predisposition is present along the crests of the main scarps and ridges of the Weser-Leine Uplands. In general, slope instability concentrates towards the top of these mountain chains, where it cumulates on large area by retracing the major relief orientation. Several clusters with high probability of landslide occurrence are also located in different sections and positions of the Weser valley and its tributaries. The spatial pattern of unstable area in this region is relatively disperse, as slope instability shows a clear fragmentation in major landslide sites and large zones with even no predisposition to mass movements. The third key area covers parts of the Harz Mountains, where significant landslide

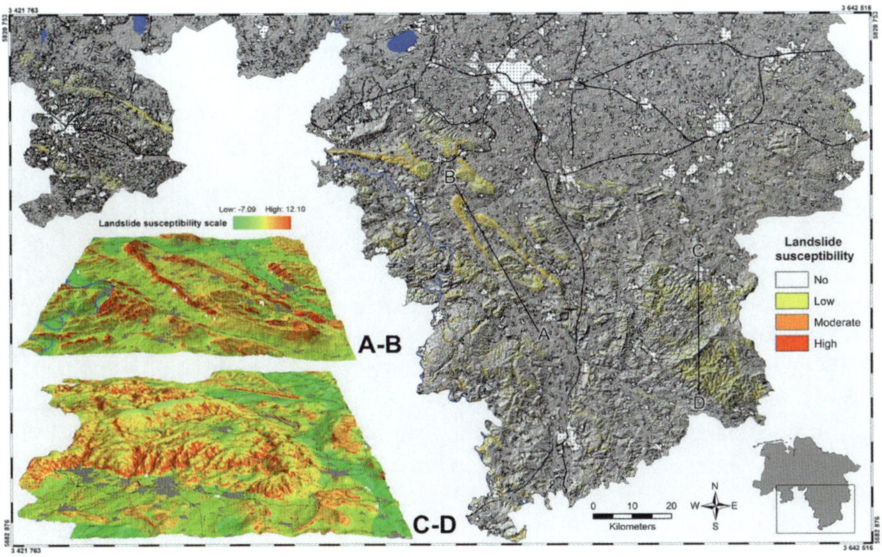

Fig. 5.1 Landslide susceptibility map for the Federal State of Lower Saxony. The model estimates that slope instability is widely restricted to the Lower Saxon Uplands illustrated in detail. Two transects exemplarily display the spatial patterns of landslide susceptibility in an area of the Weser-Leine Uplands and the Harz Mountains (*Source* Klose et al. 2014a)

Table 5.2 Scaling and spatial extension of the derived susceptibility classes (*Source* modified after Klose et al. 2014a)

	Landslide susceptibility class			
	No	Low	Moderate	High
Class scaling *Based on SI(x)*	−7.09–3.00	3.00–6.00	6.00–9.00	9.00–12.10
Spatial extension (km^2) *Based on entire Lower Saxony*	46.682	704	200	14
Spatial extension (%) *Based on entire Lower Saxony*	98.07	1.48	0.42	0.03

susceptibility is distributed on many valley sides. Nevertheless, its level is somewhat lower compared to the other regions, since it reaches only locally moderate to high values. Besides these geographic centers of slope instability, there are isolated zones of high predisposition, whose distribution, however, does not follow a clear spatial pattern.

The landslide susceptibility model estimates that about 2 % of the territory of Lower Saxony is potentially affected by mass movements (cf. Table 5.2). This corresponds to a total area of 918 km^2. A majority of this land is belonging to the lowest susceptibility class, which constitutes 77 % of the landslide-prone terrain.

Moderate predisposition is attributed to 22 % of the unstable ground, meaning that an area of 200 km^2 shows enhanced probability of landslide occurrence. In consequence, the zone of high susceptibility, in which landslides are ubiquitous and frequently recurring phenomena, extends over 14 km^2. Regarding the spatial landslide significance, it has to be considered that this areal coverage almost exclusively refers to southern Lower Saxony, which represents less than 30 % of the entire territory.

The susceptibility map reveals that urban area and transportation infrastructure is often situated in zones with an increased level of landslide predisposition. In southern Lower Saxony, about 21 km^2 of urban area shows a potential exposure to mass movements (cf. Table 5.3). Since most part of the Lower Saxon Uplands is highly developed, this area at risk merely makes up 1 % of the built environment. In this context, potential landslide exposure worth considering is not only identified for housing and industrial land, but also for recreation facilities. On the other hand, large segments of the road network pass through unstable terrain. Thus, nearly 14 km of road in the highway system of southern Lower Saxony is potentially threatened by mass movements. This corresponds to about 2 % of network's total length. Regarding state roads, this number is even higher, as there are 89 km of road, which are built on landslide-prone hillsides. In consequence, slope instability is likely to affect the state road network on up to 4 % of its coverage. All these places are identified as areas at risk, where property damage and personal injury have to be expected. Their spatial distribution is more or less diffuse, but there is a concentration of a high potential of infrastructure exposure, where land use activity penetrates in areas of high relief intensity.

Table 5.3 Potential exposure of urban area and road infrastructure to landslides in the Lower Saxon Uplands

Land use type	Spatial extension (km^2, km)	Susceptibility class	Exposure (km^2, km)	Exposure (%)
Urban area	1873	Low	17.28	0.92
		Moderate	3.66	0.20
		High	0.21	0.01
		Total	**21.15**	**1.13**
Highways	676	Low	11.94	1.77
		Moderate	1.95	0.29
		High	0.00	0.00
		Total	**13.89**	**2.06**
State roads	2265	Low	65.32	2.88
		Moderate	21.60	0.95
		High	1.80	0.08
		Total	**88.72**	**3.91**

These values base on the land use data displayed in Fig. 5.1, which is why they have to be treated as estimates (*Source* modified after Klose et al. 2014a)

5.1.2.3 Identification of Potential Infrastructure Exposure—Two Examples from Southern Lower Saxony

Using the example of the Weser Mountains and the Upper Weser region, two areas of the Lower Saxon Uplands with potentially high exposure of infrastructure to landslides are presented below in more detail. The concept of potential infrastructure exposure is understood as a likely threat for developed land through the presence of facilities and land use activity on a probable landslide mass. Since regional landslide susceptibility maps usually do not integrate run out distances, this concept cannot take into account the potential exposure of infrastructure situated in a possible path of landslide movement. Despite this technical limitation, both regions are identified to include extensive clusters of areas at risk, which are characterized by the following specifics:

(a) The Weser Mountains rise at the northern edge of the Lower Saxon Uplands about 200 m above their foreland and form an Upper Jurassic limestone ridge that contains unstable marl- and claystone sequences (cf. Hesemann 1975). Almost the entire mountain chain shows significant landslide susceptibility, which implies serious risks to the trans-European highway E 30 and a production site of the limestone mining industry (cf. Fig. 5.2a). The route of the highway traverses not only an extensive area of moderate to high predisposition, but might also be affected by adjacent landslide activity, since its location at the foot of the escarpment indicates a close proximity to highly unstable slopes. In this landslide-prone terrain, inadequate land use practices already initiated slope failure, as the example of the mineral exploitation nearby the highway E 30 illustrates. In the year 2004, the longtime undermining of the limestone ridge resulted in the collapse of the quarry face, which caused a 500,000 m^3 large rockslide. The implications were direct property damage as well as high indirect losses through business disruption, protection and monitoring measures and slope restoration (cf. NNG 2013; Meyer 2005).

(b) The Upper Weser region belongs to the Bunter Sandstone area of the Solling anticline located in the southernmost part of Lower Saxony. The focus is on the city of Hann. Münden (cf. Fig. 5.2b), where the rivers Fulda, Werra and Weser eroded a 300 m deep valley basin into Triassic sandstone formations composed of silt- and claystone interbeds (cf. Backhaus et al. 1980). Over the past decades, land use pressure had caused an expansion of the urban territory to valley slopes that are identified to be very susceptible to landslides. Today, large area of residential property and public facilities represents a zone at risk, in which landslide activity endangers infrastructure in different ways (cf. Damm and Pflum 2004). Besides severe damages through abrupt slope failure, slow creep processes show the potential to cause total losses in the long run. As a result of hillside development on unstable terrain, this community and its vicinity is faced with periodic and continuous damage to buildings and lifelines. The economic losses amount to thousands of annual damage and prevention costs and millions in periodic maintenance expenditures (cf. Damm 2000; Klose et al. 2012a).

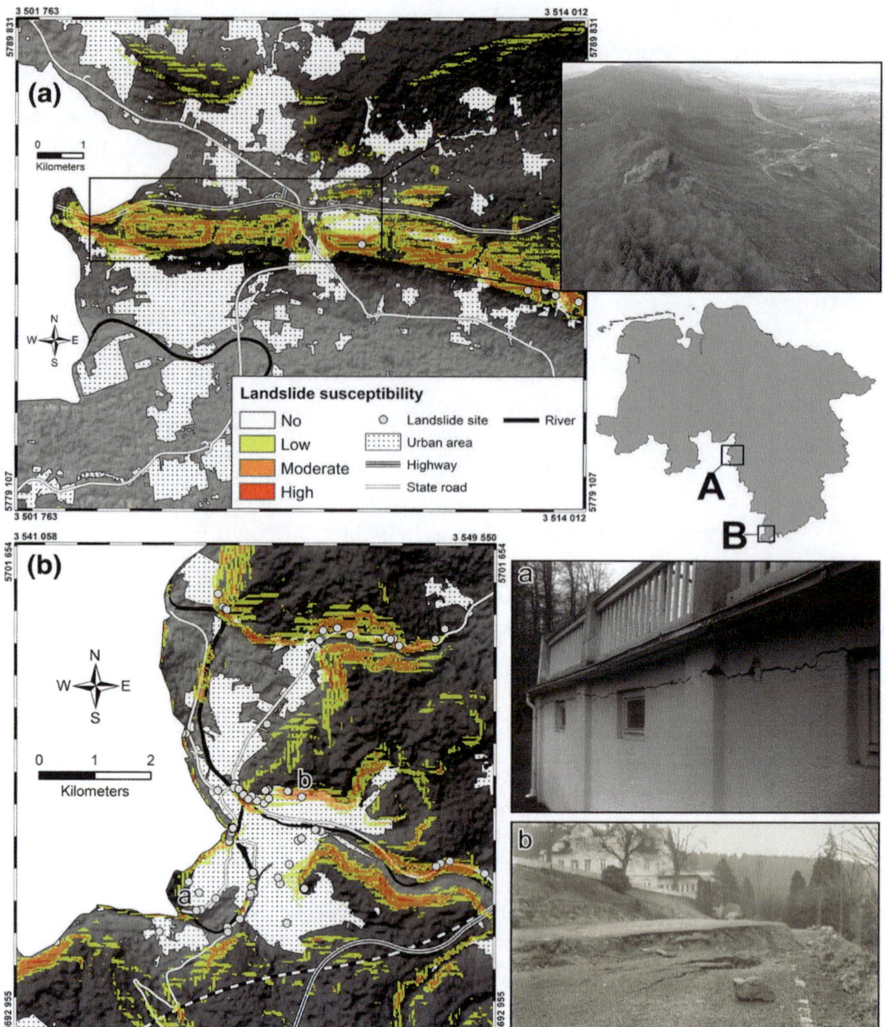

Fig. 5.2 Potential infrastructure exposure to landslides in two exemplary regions of the Lower Saxon Uplands. The case study refers to **a** the Weser Mountains and **b** the city of Hann. Münden, Upper Weser region (*Source* modified after Klose et al. 2014a)

5.1.3 Model Assessment

5.1.3.1 Validation of the Model

The validation using the success and prediction rate verifies accurate model performance and shows from technical perspective that the model is suitable for a first spatial evaluation and zonation of landslide susceptibility for overview

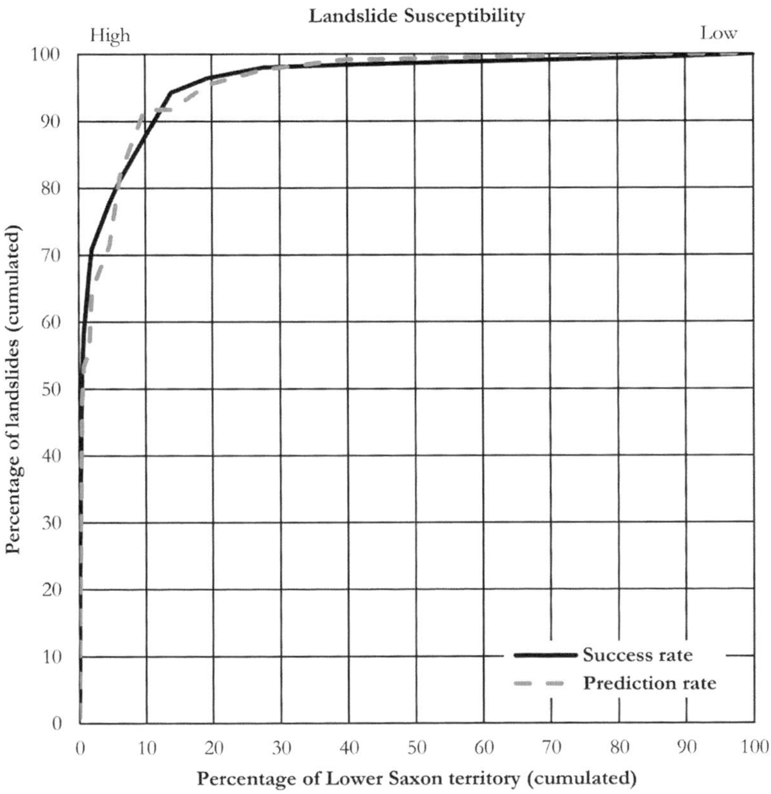

Fig. 5.3 Success and prediction rate of the susceptibility model. The success rate is derived on the basis of 756 landslides stored in the modeling set. The calculation of the prediction rate rests on 133 landslides belonging to the validation set. Both graphs prove that the model has good predictive power and high spatial accuracy (*Source* Klose et al. 2014a)

purposes (cf. Fig. 5.3). The success rate measures how precisely the model reproduces the landslides of the modeling set (cf. Chung and Fabbri 2003). The graph of this rate demonstrates that 88 % of the landslides included in the model refer to the most susceptible 10 % of the Lower Saxon territory. On the other hand, the prediction rate specifies how good the independent landslides of the validation set are predicted by the model (cf. Chung and Fabbri 2003). As the graph of this rate shows, 92 % of these mass movements are located in the most unstable 10 % of study area. In general, it holds that if a majority of landslides is concentrated in the highest 10 % interval of susceptibility, the model is proven to be reliable (cf. Chung and Fabbri 2003). Since this is the case, the accuracy of the modeling approach is confirmed, even though this result is no indicator for the model's general plausibility (cf. Sect. 5.1.4).

Table 5.4 Contingency table for slope gradient and land use

Land use	Slope gradient							
	0°–7°	7°–14°	14°–21°	21°–28°	28°–35°	35°–42°	42°–49°	Total
Forest	27	48	143	201	97	29	6	551
Grassland	12	10	12	19	9	4	0	66
Arable land	21	4	8	7	7	0	0	47
Urban area	59	14	5	10	4	0	0	92
Total	119	76	168	237	117	33	6	756
Measures of association and χ^2-Test								
$\chi^2 = 264.69$				H_0: Land use is independent from slope gradient				
$C = 0.51$				$\chi^2 > \chi^2_{0.01}$, where $\chi^2_{0.01} = 42.31$				
$C_{corr} = 0.59$, ε [0,1]				→ H_0 rejected, no conditional independence				

The measures of association and the χ^2-Test indicate statistical dependency between both geofactors (*Source* modified after Klose et al. 2014a)

5.1.3.2 Test of Conditional Independence

A few statistical procedures are available to test the conditional independence of the input data. Besides contingency statistics, the χ^2-Test is used in this study, as both test methods are widely accepted (cf. Agterberg and Cheng 2002; Pereira et al. 2012). In the present context, the problem of data dependency is reduced to the spatial association between slope gradient and land use. Different measures of association, especially C_{corr} with a value of 0.59, illustrate a significant relationship between both geofactors. This is verified by the χ^2-Test, which clearly rejects the null hypothesis H_0, indicating that conditional independence is violated (cf. Table 5.4). Nevertheless, land use is still incorporated in the model, as the observed dependency only leads to a bias of absolute susceptibility, whereas its scale order and areal delineation remains unaffected (cf. Agterberg and Cheng 2002; Neuhäuser et al. 2012a).

5.1.4 Discussion

5.1.4.1 Modeling Approach and Limitation of Input Data

GIS-based regional landslide susceptibility modeling by means of a bivariate statistical approach faces several methodological and data-related drawbacks. Basic assumption of every statistical model is the conditional independence of its input data (cf. Van Westen 2000). This prerequisite is not fulfilled, if different geofactors show comparable spatial patterns, so that it is not possible to differentiate their specific influence on instability. In case of factor dependency, landslide predisposition is overestimated at locations, where correlated geofactors spatially coincide (cf. Agterberg and Cheng 2002; Pereira et al. 2012). A significant dependency is tested to be existent between slope gradient and land use. This dependency of both geofactors is rarely addressed in literature, even though land use patterns suggest its omnipresence, especially in Central Europe. In principle, data dependency can be eliminated by exclusion or combination of geofactors (cf. Van Westen 2000). However, this is not required in the present case, as the correlation is verified to have no substantial impact on landslide zonation. It only influences the total value of susceptibility, leading to its constant overestimation, but not changing the statistical relations in the susceptibility scale. Despite data dependency, rank order and position stays the same, having no effect on the spatial delineation of landslide predisposition (cf. Sect. 5.1.3).

Many limitations in susceptibility assessment arise from the high sensitivity of statistical modeling procedures to input data quality. One important aspect refers to the completeness of the inventory in comparison to the size of the study area (cf. Thiery et al. 2007; Guzzetti et al. 2006). Serious problems are expected to occur, if the inventory is not spatially homogenous. This is the case, when the landslide record is limited and shows strong variation in its geographical coverage. Generally, there are no standards about the level of completeness, but the inventory should be as completed as possible (Van Westen et al. 2008). In this investigation, most landslides belong to the Weser-Leine Uplands, whereas the Harz Mountains are under-represented in the inventory (cf. Fig. 4.3). Nevertheless, this is proven to have no critical effect on the identification of susceptible areas, even though it results in a regionally too conservative evaluation. An important problem of statistical models emerges from their spatial scope itself. The regional perspective implies considerable data limitation due to reduced availability as well as labor and cost intensive acquisition. Therefore, it is common practice to develop simplified models, which are based on input data that are easy to collect, but often imply further methodological problems (cf. Thiery et al. 2007; Van Westen et al. 2008).

In the present study, this is the case with lithology, as a medium to large-scale geologic map is too precise in its representation. A high level of detail implicates many lithologic classes showing low landslide frequency, so that useful correlations are hardly able to reveal. Most suitable is a generalization of the Lower Saxon geologic overview map 1:500,000, although it has the drawback that some bedrock classes are not mutually exclusive (i.e. sand-/limestone, clay-/limestone). This conflicts with conditional independence, but has to be accepted, since no alternative data sources are available. A further drawback is that this geologic map ignores near-surface subsoil, where landslide activity is frequently taking place, as field survey and previous research work indicate (cf. Sect. 3.2). In consequence, correlation with rock type must be seen critically, as predisposition is rather determined by soil than by bedrock properties. Susceptibility models usually have a static character, so that land use, which shows high dynamics in space and time, can only be poorly represented (cf. Van Beek and Van Asch 2004; Van Westen et al. 2008). The problem associated with land use is not just its changing nature, but the difficulty to explain its causalities using a model, which solely focusses on the spatial presence or absence of certain geofactors. In this sense, it is questionable, whether the influence of urban land use can be related to the factor`s pure spatial existence. Up to now, this problem is widely neglected, although it is of high relevance. An advantage of regional studies is the good access to cost-free DEMs, offering useful relief information, but only in resolutions that conflict even with large-scale purposes. In this study, the ASTER GDEM shows general applicability, yet accompanying deficits in its spatial accuracy decline model quality. This expresses in underestimating slope gradient and the ignorance of small-scale terrain variation. The use of high-resolution DEMs provides a solution, but poor availability and high costs are still problematic, forcing the application of ASTER imagery in the present case.

5.1.4.2 Data Processing and GIS-Based Modeling

Different aspects concerning the data processing and the statistical modeling using GIS are worthwhile to comment in more detail. A major point of criticism relates to the spatial accuracy of the data transfer from the landslide map of Schunke (1971). This inventory illustrates landslides at a scale of 1:200,000 in terms of simple point and polygon symbols. Despite of careful data digitization and georeferencing, the import of the landslide information is still associated with some imprecision. An exact localization is even with additional DEM data and satellite imagery difficult to achieve. In view of this, errors in geopositioning are likely to occur, whereby their propagation may result in model inaccuracy. Another problem is the spatial resolution of the ASTER GDEM, which does not only hamper landslide mapping, but also affects the computation and accuracy of the derived slope parameters. Their precision generally depends on the cell size of the applied DEM. As result of view and image geometry, slope gradient derived from a DEM is negatively correlated to the size of its grid cells, implicating a lack of exactness in topographic representation, if applying a DEM with low to medium resolution (cf. Zhang et al. 1999). Several methodologies have been proposed to address this problem by transforming and re-scaling of obtained slope data (cf. Qinke et al. 2008). In landslide susceptibility modeling, this type of error gains no high attention so far, but is yet of relevance, as cross-checking with ground truth data exemplifies. By comparing slope data from the landslide database with that of the DEM analysis, it becomes obvious that there is a slight systematic bias towards underestimating slope gradient. This primarily affects steep slopes in areas of high terrain variation, where negative deviation of up to 5°–10° is verified at certain landslide sites. Technical solving of this error, however, is not of top priority, but accuracy concerns require keeping in mind the consequent model deficiency.

A modified information value approach is at the core of the present susceptibility model. The modification concerns the formulation of the weighting function and intends to simplify both data processing and interpretation, so as to meet the specific requirements on regional level. Major difference to previous conceptualizations is the way how landslide information enters the calculation, which means that landslide pixel mapping is substituted by point representation and the weighting function uses landslide densities based on attribute areal coverage. Usually, pixel-based procedures dominate, deriving landslide density from all grid cells defining slope instability. The delineation of extensive landslide sites by pixel mapping, however, turns out to be sophisticated in studies with regional focus. This is why an alternative approach has been developed. According to this concept, each landslide is localized using a single data point, which reduces spatial landslide extent to the area of one grid cell. As a result, inaccuracy increases due to restricting input data, but the advantages through easier data processing are worth the loss in information, which can be minimized by proper geocoding (cf. Sect. 4.2.2). Nevertheless, there are problems related to the representation of extensive landslide areas, which are geocoded by taking the spatial reference of a specific mass movement feature. In this study, ground truth data and the application of Google Earth® assisted landslide localization in few cases with imprecise coordinates, as ASTER imagery is of too low accuracy for this. Subsequently, landslide density is derived by putting the number of landslides per attribute class in relation to the areal coverage of the respective attribute. In contrast to a pixel-based approach, the obtained density value clearly reveals the spatial landslide significance, since providing easily interpretable data on the amount of landslides per km^2. In conclusion, this modification reduces model precision, but is an effective way to enable susceptibility modeling given the regional specifics.

5.1.4.3 Modeling Results and Susceptibility Map

Landslide Controlling Factors and their Mechanisms

Slope gradient is identified to be the most relevant factor controlling landslide susceptibility in Lower Saxony. This corresponds not only with findings of regional detail studies (cf. Damm et al. 2009; Varga et al. 2006), but also matches with modeling results worldwide (cf. Van Den Eeckhaut et al. 2006; Dai and Lee 2002; Ohlmacher and Davis 2003). In agreement with other investigations, however, the results indicate that even slope gradients well below 30° can be related to significant landslide predisposition (cf. Bălteanu et al. 2010; Neuhäuser et al. 2012b). Therefore, it can be concluded that critical slope gradients show a close dependency to lithologic properties. The model reveals that slope curvature is in terms of strong concavity or convexity of second highest influence on landslide occurrence. These findings conflict with field evidence, which outline that primarily concave slopes tend to instability, as their shape favors moisture anomaly caused by runoff concentration (cf. Sidle and Ochiai 2006). Nevertheless, some studies show this kind of inconsistency (cf. Ayalew et al. 2004; Conforti et al. 2012; Havenith et al. 2006), which means that the spatial distribution of landslides is not strictly characterized by this causality. Slope aspect, in contrast, does not have a substantial effect on landslide susceptibility. Slopes of all main directions are affected by instability, although a slight dominance of NE sector is observed.

This spatial pattern is more likely the result of the orographic orientation (cf. Fig. 5.1) than of enhanced moisture retention on shady slopes, as few investigations highlight (cf. Dai and Lee 2002). Furthermore, a spatial relationship with the regional distribution of rainfall, on which some studies report (cf. Komac 2012; Neuhäuser et al. 2012a; Van Den Eeckhaut et al. 2006), cannot be found for Lower Saxony, despite comparing landslide patterns with that shown in different precipitation maps not considered in the modeling.

The influence of lithology on slope instability needs to be interpreted with caution, as both data quality and inventory homogeneity are limited. However, there is clear evidence that landslides are closely connected to Mesozoic sedimentary rock. According to previous studies (cf. Ackermann 1953; Damm 2005; Schunke 1971; Tilch 1999), the findings reinforce the high susceptibility of sand- and limestone formations composed of weak clay- and marlstone interbeds. Due to the limitation of the lithologic input data, no direct conclusion on the significance of hillslope sediments can be drawn, although other investigations suppose their high relevance in Lower Saxony (cf. Damm et al. 2010; Klose et al. 2012b). The results concerning the role of land use underline the inconsistency existing in literature. In many cases, land use is evaluated to be of different influence on landslide initiation, whereby data inconsistency is rarely addressed specifically (cf. Magliulo et al. 2008; Van Den Eeckhaut et al. 2012). Thus, the positive landslide susceptibility of forest areas found in this study is more likely related to the dependency of land use patterns on geomorphometry than to a doubtful destabilizing effect of forest vegetation. Altogether, land use is of lower priority for slope instability, which also matches with other studies (cf. Conforti et al. 2012; Wang and Sassa 2005).

On regional level, it finally raises the question, if it is possible to achieve plausible modeling results using only slope gradient as input variable. Such an approach simplifies the investigation process and has already been applied in other studies (cf. Godt et al. 2012). The validation of a respective model proves its capability to perform a reliable landslide prediction for Lower Saxony. Thus, the prototype map reproduces the susceptibility model with almost 80 % accuracy, as the comparison of both success rates shows.

Spatial Landslide Susceptibility Pattern

The study reveals that the regional setting of landslide susceptibility is like in other European regions or in some countries worldwide highly correlated with the location and composition of the mayor relief structures (cf. Nadim et al. 2006; Van Den Eeckhaut et al. 2012). Landslide susceptibility has a clear spatial pattern and is widely restricted to the mountainous south of Lower Saxony (cf. Fig. 5.1). About 2 % of the Lower Saxon territory is landslide-prone, which is comparatively low in the European context (cf. Bălteanu et al. 2010; Jelínek et al. 2001; Trigila and Iadanza 2008), even though direct comparison is hardly plausible. Unstable terrain exists on large area especially in the Weser-Leine Uplands, the Upper Weser region and the Harz Mountains. The analysis of the constellation of susceptibility in these main distribution areas demonstrates its spatial association to specific relief positions and landforms. Typical for the cuesta landscape of the Weser-Leine Uplands is the concentration of high landslide predisposition at the top of the main scarps and ridges. In the Weser Valley, major clusters of potential slope instability are related to cut banks, narrows and steep basins. Landslide susceptibility is widespread on the valley flanks of the Harz Mountains, but reaches only in steeper sections of the V-shape valleys a significant level. Furthermore, the distribution of landslide-prone area is characterized by isolated clusters, which are difficult to assign to a broader topographic setting. They correspond to some extent with single hills, scarps, terraces and artificial slopes. The findings on the relationship of landslides to certain landforms and relief structures may help to assess slope instability in comparable landscapes of Central Europe.

5.1.4.4 Application of Regional Landslide Susceptibility Maps to Assess Potential Infrastructure Exposure

A number of reports and guidelines address the creation of landslide susceptibility maps for the purpose of land use planning (cf. Fell et al. 2008; Schwab et al. 2005). Nevertheless, there are almost no studies that comment on their application to assess potential infrastructure exposure (cf. Guillard and Zezere 2012). One major objective of regional landslide susceptibility maps is to brief policy makers and the general public about the potential occurrence and distribution of landslide hazards (cf. Fell et al. 2008). More specifically, they must be seen as a first tool for the pre-selection and delineation of regional landslide priority areas, so as to provide a basis for further local investigations, focusing on efficient site investigation and selection. On the other hand, their function in regional hazard management is to give an overview of the potential exposure of communities and infrastructures to landslides on subordinate level, which should encourage further detail studies to develop local protection strategies (cf. Schuster and Highland 2007). However, most regional landslide susceptibility maps do not fulfill these tasks, as they usually ignore the broad spatial perspective on the conflict between land use and landslide activity.

In contrast to previous studies, this point of view is addressed in the present investigation. The analysis points out that in the Lower Saxon Uplands more than 1 % of the urban area and up to 4 % of the road network is found in landslide-prone area. Until now, no regional benchmarks are available, which makes it hardly possible to evaluate these numbers properly. Only Guillard and Zezere (2012) provide comparable reference figures that amount to 2 % for urban area and up to 10 % for different road categories. In southern Lower Saxony, the spatial extent of potential infrastructure exposure varies between the different key areas of slope instability. Thus, the Harz Mountains are widely free of urban area at risk, but show a potential of low to medium exposure along some major state roads. However, landslides pose a more extensive threat to infrastructure in the denser populated Weser-Leine Uplands. The spatial pattern of endangered settlement is more diffuse in this region, albeit a clustering of potential infrastructure exposure can be observed in areas, where spatial development is forced to take place on steep hillsides. These conditions exist especially in the Weser Mountains and the

Upper Weser region, in which the problems of developing unstable land are clearly evident. Both case studies show that there is a need for information on regional landslide susceptibility.

5.2 Temporal Hazard—When and Why Do Landslides Occur?

Note: The entire section has been published in Klose et al. (2012b).

5.2.1 Stability Criteria and Strength Properties of Hillslope Sediments

The mantle rock as lowermost unit of the three-part cover bed complex is in general compara-tively stable. This is primarily due to a high coarse soil proportion, a high proportion of grains >630 mm and low silt and clay proportions in the fine soil. Findings indicate that also colluvial deposits largely possess stability as result of the internal friction of the substrate and slope inclinations of mostly <25°. Both units are, therefore, only of subordinate relevance in any slope stability considerations (cf. Table 5.5). Landslides, in contrast, are mostly developed in Fließerde, whose strength properties are considerably lower. These sediments, in particular slightly plastic silts and sand-silt-mixtures, are largely unstable, because of their specific grain size distribution and their sensitivity to water supply. The cohesive fine soil responds quickly to moisture penetration with changes in consistency and a reduction of soil strength. Even with low water absorption, consistencies turn from at least semi-compact to plastic or viscous conditions. The natural water content W ranges between 16.3 and 30.6 %, and varies depending on grain size category and proximity to the slope water level. The values of W exceed these of

Table 5.5 Characteristic soil-mechanical values for Quaternary cover beds overlaying the Middle Lower Triassic Bunter Sandstone (sm) in Northern Hesse and Southern Lower Saxony

Parameter	Layer 1 (Mantle rock)	Layer 2 (Fließerde, average condition)	Layer 2 (Fließerde, water-saturated)	Layer 3 (Soil sediment)
Specific weight (kN/m^3)	21	20.5	11	20
Friction angle (°)	35	27.5	27.5	22
Cohesion (kN/m^2)	50	5–10	0	5–10

Data derived from laboratory analysis and mean values according to DIN 1055 (*Source* modified after Klose et al. 2012b)

Table 5.6 Characteristic soil-physical parameters and Atterberg limits of Fließerden at landslide sites in Northern Hesse and Southern Lower Saxony

Characteristic value	Symbol	Min. value	Max. value	Property
Natural water content	W (%)	16.3	30.6	Often above average
Water permeability	k (m/s)	1.0×10^{-6}	1.5×10^{-6}	Semi-permeable
Plastic limit	W_P (%)	17.1	29.1	Fast change of consistency
Liquid limit	W_L (%)	19.9	32.3	Fast change of consistency
Plasticity index	I_P	0.5	5.2	Low
Consistency limit	I_C	0.39	0.58	Very low
Effective bulk density	Ld	3	–	Medium
Field capacity	FC (vol.%)	26	36.5	–
Usable field capacity	EFC (vol.%)	15	22	–
Frost sensitivity	F	F3	–	Very frost sensitive

Data derived from 34 samples according to DIN 18122 (*Source* modified after Klose et al. 2012b)

equivalent soils, and are close to the flow limit. In addition to intermediate bulk densities, water permeability values are semi-permeable, so that these sediments act more or less as water-retaining stratum. Shear strength parameters are material characteristic, and can be classified as low. Finally, very low frost sensitivity supports soil fabric disintegration. Thus, the Fließerde shows high landslide susceptibility, and tends to develop landslide types in transition between sliding and flowing (cf. Tables 5.5 and 5.6).

5.2.2 Temporal Development of Soil Water Balance

The development of *SWB* during the past 59 years is characterized by an alternation of different soil moisture conditions (cf. Fig. 5.4). While perennial wet and dry periods succeeded in the 1950s and 1960s, most of the 1970s were particularly dry. The 1980s, the 1990s and the early 2000s showed strong seasonal fluctuations with partly very wet winters, but also distinct dry conditions in summer. Since 2005, the study area faces a long-lasting wet period prevailing in both seasons. The hygric differentiation in positive and negative wetness anomalies bases on the mean values of *SWB* that are calculated for the hydrological summer and winter. Despite of an increase in precipitation of around 6 % since 1953, soil moisture does not rise significantly throughout the whole period. The increase in precipitation can be attributed to a surplus of about 5 % after 1978, which, however, was largely cancelled by higher evapotranspiration, so that soil moisture remained almost constant over the total time horizon.

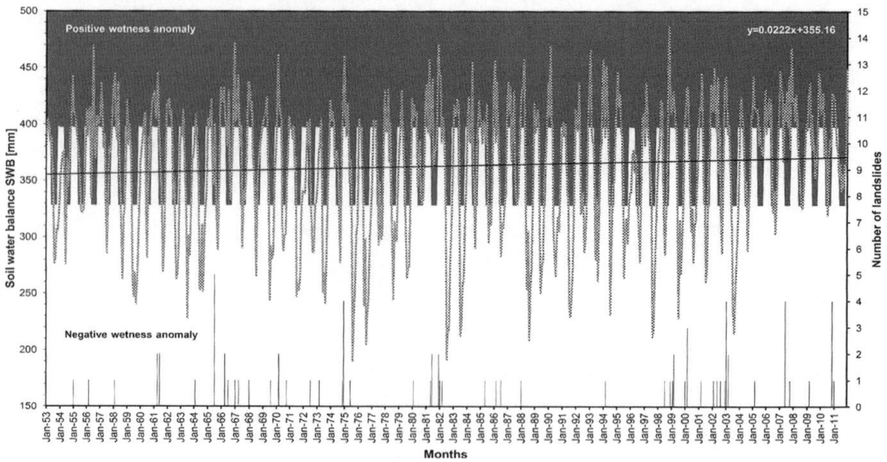

Fig. 5.4 Relationship between landslide activity (i.e. number of landslides) and simulated soil water balance (*SWB*) in Northern Hesse and Southern Lower Saxony from January 1953 to December 2011. The hygric specification (i.e. positive or negative wetness anomaly) refers to the mean value of *SWB* for hydrological summer and winter (*Source* Klose et al. 2012b)

5.2.3 Correlation Between Landslide Activity and Simulated Soil Moisture

The simulated soil moisture development shows high temporal correlation with the reconstructed landslide activity. The mass movement clusters between 1955 and the early 1970s coincide such as those at the beginning of the 1980s and these between 1998 and 2003 with periods of high soil moisture conditions. In contrast, there are no or only few landslide events in periods that are identified as dry. This holds especially for the late 1970s as well as the early and mid-1990s. The outlined development reflects the temporally variable stability conditions in hillslope sediments, which depend mainly on different soil moisture levels. Some periods depict a clear relationship between enhanced landslide activity and a strong increase of *SWB* after long-lasting dry phases. This effect, for instance, is observable at the end of 1974. From this, it can be concluded that mass movements do not only occur after or during prolonging wet phases, but also when high values of *SWB* follow a distinct dry period. Landslides are accordingly initiated by moisture impulses, which, however, can also trigger mass movements during periods of higher soil moisture (cf. Fig. 5.5).

5.2.4 Critical Soil Moisture Threshold

Soil moisture thresholds are of particular importance for process analysis, and play a key role in hazard prediction. In the study area, as the distribution of *SWB* per landslide event clearly illustrates, mass movements predominantly occur in

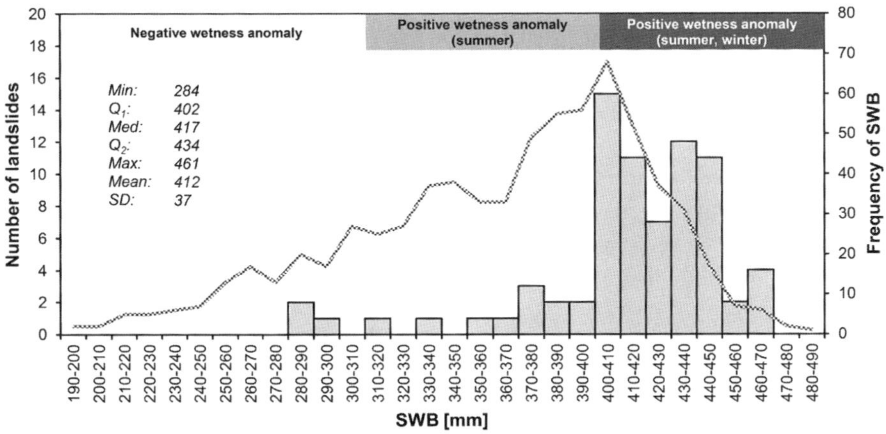

Fig. 5.5 Frequency of landslides (bar chart) and monthly soil water balance (*SWB*) for the time period January 1953 to December 2011. Both graphs and the statistical measures of the distribution *SWB* per landslide event show a close dependency of mass movements to the minimum threshold $SWB_{thres} \geq 400$ mm. The hygric specification (i.e. positive or negative wetness anomaly) refers to the mean value of *SWB* for hydrological summer and winter (*Source* Klose et al. 2012b)

months, which, in any case, can be classified as wet. Landslide initiation shows close dependency to high soil moisture levels. About 75 % of the recorded landslides result at $SWB \geq 400$ mm. In contrast, only 14 mass movements refer to soil moisture levels ≤ 400 mm, whereas at least 7 of them are demonstrably linked to high-intensity rainfall in summer. However, due to the monthly temporal resolution, daily peaks of soil moisture levels are not detected, so that storm-triggered landslides cannot be explained by the underlying conceptual framework. For this reason, $SWB = 400$ mm can be outlined even more clearly as minimum threshold SWB_{thres}, since, in our context, mass movements usually do not occur below this value. Furthermore, great importance can be attached to this threshold, as its exceeding causes a volatile increase in the number of slope failures. A total of 15 mass movements are allotted to the class 400–410 mm, which therewith possesses 19 % of all landslide events, and thus has the highest class frequency. Soil moisture classes in the range of 400–450 mm apparently constitute the central and clearly definable part of the distribution, on which more than 70 % of the mass movements concentrate on. Due to this concentration of values, statistical dispersion is quite low, as indicated by $SD = 37$ mm, although few negative outliers account for a comparatively large range of $R = 177$ mm (cf. Fig. 5.5).

5.2.5 Recurrence Frequency

Statistical analyses prove that critical soil moisture levels have partly very high recurrence frequency. Thus, the soil moisture class 400–410 mm shows not only the highest number of landslide events, but also refers, as mode of the respective distribution, to 68 months or to 10 % of the total time series, respectively. Class frequencies yet markedly drop down above the mode class, while they level more moderately below to it. During the past 59 years, 31 % of all months exceeded SWB_{thres}. The distribution of recurrence frequency consequently describes a U-shaped relationship (cf. Fig. 5.5). Annualities in the range of up to several decades are characteristic for soil moisture classes <300 mm and >470 mm. By contrast, the classes in between have return periods that, in most cases, are lower than 2 years. In the study area, soil moisture levels, at which landslides are generally possible, must be anticipated every year. However, this does not mean that mass movements are expected to occur annually, since months with landslide activity, especially in classes slightly above SWB_{thres}, have only medium recurrence frequency. If *SWB* reaches a critical soil moisture class, the probability of at least one landslide ranges from 10 to 47 %. The intuition that conditional probabilities increase with higher soil moisture classes can only be partly stochastically confirmed (cf. Table 5.7).

5.2.6 Landslide Volume and Soil Moisture Level

Beside the influence of positive wetness anomalies on landslide activity, the investigation gives first evidence about an existing correlation between landslide volumes and soil moisture levels. Data evaluation demonstrates that small landslides with volumes of $V \leq 100$ m^3 occur over the entire spectrum of critical *SWB*. In contrast, large mass movements ($V > 1.000$ m^3) are generally bound to soil moisture levels >410 mm. In the study area, soil moisture conditions supporting large slope failures recur in periods of at least 1.1 years (cf. Table 5.7).

Table 5.7 Recurrence frequency and landslide probability of relevant soil moisture classes (*Source* modified after Klose et al. 2012b)

SWB (mm)	Recurrence frequency (years)	Landslide probability (%)
190–200 and 200–210	29.5	–
210–220 to 290–300	≤11.8	*Not considered*
300–310 to 390–400	≤2.4	*Not considered*
400–410	0.9	10.3
410–420	1.1	17.3
420–430	1.6	10.8
430–440	1.9	22.6
440–450	3.5	47.1
450–460	8.4	28.6
460–470	9.8	16.7
470–480	29.5	–
480–490	59	–

5.2.7 Period of Saturation and Initiation Time

Apart from the positive correlation between landslides and soil water surpluses in the month of occurrence, the duration of the previous wet phase is assumed to be a crucial pre-disposing factor. When precipitation exceeds evapotranspiration, the soil saturates as long as SWB_{thres} is reached. The saturation time generally depends on the saturation state of the soil, and has thus a different duration. Once SWB_{thres} is reached, the initiation time is determined by how long and to which extent the critical value is exceeded. In this context, the time of exceeding SWB_{thres}, the duration of the previous wet phase and the number months that outreach FC were analyzed (cf. Fig. 5.6). Referring to SWB_{thres}, it can be pointed out that initiation times generally range from 0 to 8 months. About 60 % of all mass movements are connected with an initiation time $\geq$1 month. In view of the wet phase length and the exceeding time of FC, an even more clear relationship arises, as both variables highlight that approximately 75 % of the recorded landslides are related to positive initiation times. This especially holds for the wet phase length, which occasionally lasts for up to 16 months. However, 86 % of all mass movements have a wet phase length $\leq$6 months. Additional statistical analyses prove the trend that landslide initiation requires in case of low initiation time considerably higher soil moisture levels than in high initiation time. By contrast, there is yet no clear relationship between initiation time and landslide volume. Large landslides can occur just as small ones at any time when SWB_{thres} is exceeded.

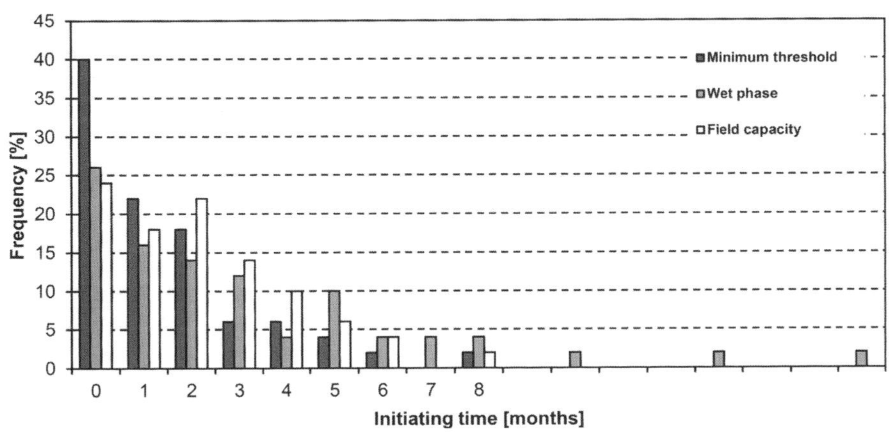

Fig. 5.6 Frequency of different landslide initiation times. The distribution refers (*1*) to the number of successive months with soil water balance $SWB \geq$ minimum threshold SWB_{thres} of about 400 mm, (*2*) to the number of successive months with $SWB \geq$ field capacity FC and (*3*) to the duration of the proceeding wet phase (*Source* Klose et al. 2012b)

5.2.8 Discussion

Beside the soil-physical and soil-mechanical parameters, the hydrological conditions play an important role for landslide initiation in hillslope sediments of the study area. Depending on precipitation and evapotranspiration, the soil moisture level is a pre-disposing and triggering factor of particular interest. This especially holds for the Fließerde, in which stagnant or flowing soil water reduces slope stability due to increasing pore water pressure and the decrease of specific weight and cohesion. This generally causes mass movements in transition between sliding and flowing. The values of *SWB*, which are calculated on the basis of COSIMO, enable direct estimation of the soil saturation level. Due to the use of monthly mean values as input variables, daily peaks of precipitation, as they can occur in the course of high-intensity rainfall, are neglected in the simulation. The calculations performed in this investigation reflect, therefore, the long-term soil water development in the study area, which is smoothed compared to daily extremes.

The temporal development of *SWB* shows good agreement with the landslide activity, and demonstrates the temporal variability of the stability conditions in hillslope sediments quite well. Statistical analyses explicitly highlight that mass movements occur almost exclusively during months with values of *SWB* $\geq$400 mm. This value is identified, therefore, as minimum threshold, whose exceeding is usually an indispensable requirement for landslide initiation. Critical soil moisture levels are proved to have partly very high recurrence frequency, and thus possess to some extent comparatively low annualities. The conditional probabilities for slope failure in critical soil moisture classes amount to 10–47 %. Contrary to small landslides that could also occur at low values of *SWB*, large ones are demonstrably linked to soil moisture levels >410 mm. An up to 16 month long initiation time with positive wetness anomalies usually proceeds the month of the slope failure. However, it is important to note that mass movements can be also observed when high soil moisture levels follow a pronounced dry phase. But such moisture impulses could also be causative during prolonging wet phases, since the initiation time depends, amongst others, on the magnitude of exceeding a respective threshold.

The results indicate that a monthly based approach possesses enough explanatory power for the present purpose. In contrast to previous studies, which primarily have daily perspectives, it offers the advantage of comparatively low data requirement. Therefore, the model helps to assess hazard potentials over longer time periods. With regard to current research, no adequate importance is attached to the relationship between landslide activity and long-term soil moisture development. Hardly any studies can be used to compare the obtained results. However, referring to existent antecedent rainfall thresholds (see above), research findings show considerable differences. Although high initiation times were partly proved already, insights from Northern Hesse and Southern Lower Saxony yet illustrate that the relevance of proceeding wet phases is underestimated so far.

Notwithstanding the partly very good correlation results, future research work has to concentrate on combining the long-term soil moisture development with daily rainfall data directly before landslide occurrence. Herewith, the existing difficulties, which lie in an insufficient temporal resolution to explain landslides triggered by high-intensity rainfall, can be solved. Several studies account successfully for such a perspective (cf. Crozier 1999; Glade 2000; Godt et al. 2006; Ponziani et al. 2012), but, as mentioned above, do usually not refer to long-term trends and patterns of landslide initiating soil moisture conditions. Therefore, it is intended to perform correlation analyses, which will seize this problem in future times.

5.3 Hazard Impact—How Much Do Landslides Cost?

Note: The entire section has been published in Klose et al. (2014b).

5.3.1 Landslide Losses for Highways in the Upper Weser Area

5.3.1.1 Cost Structure and Temporal Patterns of Landslide Losses

The total landslide loss for highways in the Upper Weser area amounts to US$23.5 million between 1980 and 2010. Figure 5.7a illustrates that 19 (61 %) years of this 31-year period show landslide damage. The distribution of the annual number of landslide damage events is relatively homogenous, ranging between zero and three events per year. By contrast, the costs of landslide damage are strongly concentrated on five years in the early 1980s and the mid-2000s which together account for 94 % (US$22.2 million) of the total loss. As a result, there is huge discrepancy between the annual average costs (US$0.76 million) and the annual median costs (US$17,000), a fact of high importance for cost extrapolation (Table 5.8). Years with minimum costs of at least US$0.1 million have a return period of $T = 3.1$ years. The highest annual costs (US$7.4 million) are estimated for the year 2006 when two major projects of landslide repair and mitigation had been realized, including stabilization of a failed cut slope (US$4.6 million) and slope reinforcement by soil nailing (US$2.8 million). The analysis proves that landslide repair (US$9.8 million) and mitigation (US$13.0 million) make up 97 % of the overall costs. Most of the remaining costs are due to maintenance (US$0.6 million), while the costs of first response (US$0.1 million) are widely negligible (Table 5.8). The total annual costs and the number of landslide damage events per year are only weakly correlated (Fig. 5.7c). Although costs in years with only one landslide damage event are always below US$0.5 million, there is no clear relationship that the more landslide damage events, the higher the annual costs. As Fig. 5.7b indicates, the major cost drivers are few exceptional landslide damage events causing expensive repair or mitigation. Thus, about 75 % of the total

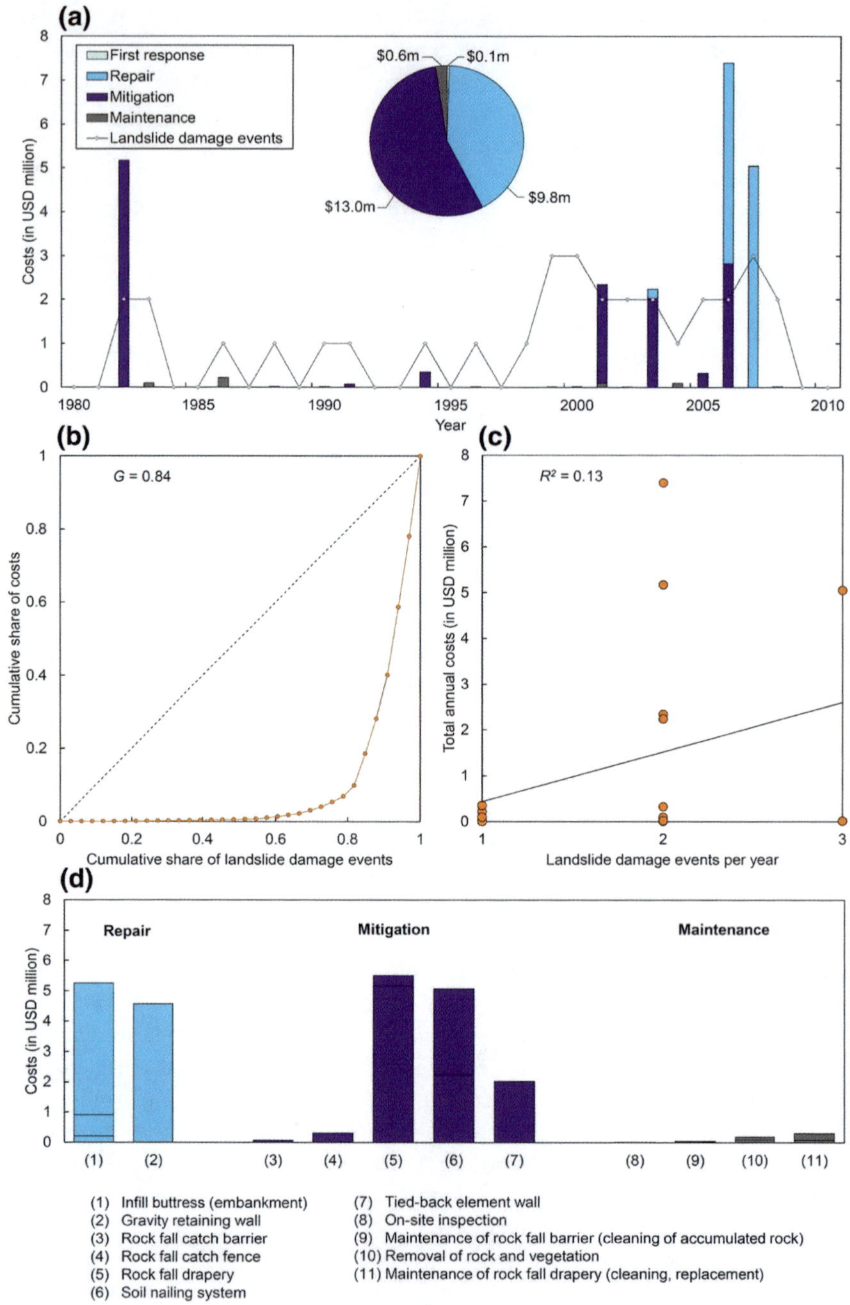

Fig. 5.7 Landslide losses for highways in the Upper Weser area between 1980 and 2010. **a** Trends of landslide activity and costs and cost structure of landslide disaster management. **b** Lorenz curve and Gini index of the landslide cost distribution. **c** Scatter plot and regression curve of total annual costs in relation to number of landslide damage events per year. **d** Types and cost structure of repair, mitigation, and maintenance measures (*Source* Klose et al. 2014b)

Table 5.8 Landslide losses and cost statistics for different types of landslide disaster management (*Source* modified after Klose et al. 2014b)

Types of landslide disaster management	Landslide damage events		Costs per damage event (in USD thousand)				Total costs (in USD thousand)	
	Total	%	Min	Max	Average	Median	Total	%
All categories	33	100	0.5	5,161	712	18	23,498	100
First response	8	24.2	6	19	–	–	101	0.4
Repair	4	12.1	208	4,571	–	–	9,822	41.8
Mitigation	7	21.2	76	5,161	–	–	13,012	55.4
Maintenance	14	42.4	0.5	222	–	–	564	2.4

costs refer to the five most costly landslide damage events. This implicates that 85 % of the landslide damage events only represent 19 % of the overall costs. The importance of repair and mitigation costs is associated with strong cost variability among the different types of landslide repair and mitigation (Fig. 5.7d). A significant part of this cost variability originates from landslide dimension and cost differences between comparable mitigation measures. This especially applies to rock fall protection by means of catch barriers and rock fall drapery that are less expensive than their counterparts (e.g., catch fence, anchored mesh systems). Such low-cost mitigation structures are identified of only temporary effectiveness, implicating short repair cycles, and thus being the major driver of maintenance costs. Among landslide damage events >US$1.0 million, a majority (67 %) is classified to cause mitigation costs. These investments in traffic safety are often a direct reaction to periods of increased landslide activity, which illustrates the high relevance of risk awareness as cost factor.

5.3.1.2 Validation of the Cost Compilation

The validation of the cost compilation for the Upper Weser area only concerns those losses gathered by cost modeling. This is because cost survey provides in principle the actual costs of landslide damage. The main idea of the validation is to cross-check the results of cost modeling with available reference data from cost survey. As the developed LDMMs describe prevailing disaster management practices, and thus show a high a degree of reliability, the validation is exclusively focused on assessing the quality of applied cost modules. To test their plausibility and accuracy, some cost modules are used to recalculate the losses of repair or mitigation measures evaluated by cost survey. The error between estimated and actual costs is seen as a first indicator for the precision of these cost modules. Such a validation was conducted for six (18 %) landslide damage events included in the cost compilation. According to Table 5.9, the error is between −11.6 and 18.7 % ($\sigma = 9.1$ %), which is within the range of tolerance of ± 10–20 % commonly accepted in project cost estimation. There are two main reasons for the cost difference between estimated and actual costs: (a) uncertainty about major cost drivers (e.g., standardized length of soil nails;

Table 5.9 Validation of cost modeling by comparing estimated costs of applied cost modules with actual costs of cost survey for different types of landslide repair and mitigation (*Source* modified after Klose et al. 2014b)

Repair or mitigation measure and relevant cost module	Area/length	Actual costs/estimated costs (in USD million)		Error (in %)
(1) Slope reinforcement after rock/soil slide, above road (year 2001) Cost module: Soil nailing, deep slip surface (12 m)	4,300 m^2	2.16	1.91	−11.6
(2) Slope reinforcement after rock/soil slide, above road (year 2006) Cost module: Soil nailing, medium-deep slip surface (6 m)	8,300 m^2	2.48	2.65	6.5
(3) Failure of fill slope, highway embankment (year 2003) Cost module: Infill buttress, medium embankment height (6 m)	20 m	0.18	0.17	−1.4
(4) Landslide in fill slope, highway embankment (year 2007) Cost module: Infill buttress, medium embankment height (6 m)	75 m	0.64	0.65	1.5
(5) Settlement of fill slope, highway embankment (year 2007) Cost module: Infill buttress, medium embankment height (6 m)	550 m	4.06	4.82	18.7
(6) Rock fall protection after small rock fall, above road (year 2005) Cost module: Catch fence, low energy absorption capacity (< 100 kJ)	470 m	0.26	0.27	3.6

case 1 of Table 5.9) and (b) decreasing unit and average costs (e.g., variable length of repaired road; cases 3, 4, and 5 of Table 5.9). For example, the decrease in average costs from cases 3 to 5 of Table 5.9 is about US$2,000, a difference in costs explaining the error of 18.7 % for case 5. The validation verifies sufficient reliability and accuracy of the tested cost modules but requires more empirical reference data for assessing the overall quality of cost modeling.

5.3.2 *Landslide Losses for Highways in the Lower Saxon Uplands*

5.3.2.1 Highway Exposure to Landslides and Regional Cost Estimate

The specific setup of the regional landslide susceptibility model leads to the result that slope gradient is the major controlling factor of slope instability in the Lower

Saxon Uplands (Table 5.1). Critical slope gradients are identified to lie between 21° and 49°. The landslide susceptibility model predicts that slopes with a high degree of plane curvature show substantial tendency to landslides. Alternatively, slope aspect is proven to be a controlling factor of only subordinate relevance. Most landslides are related to Mesozoic sedimentary rock, especially sand and limestone or clay and limestone, which exert strong positive effect on landslide occurrence. By contrast, land use shows low importance for regional landslide susceptibility; however, this ignorance of human impact is mainly due to the required model simplicity in applications on large spatial scales (cf. Klose et al. 2014a). The expert-based scaling of the susceptibility index results in 7.5 or 10.1 % of the territory of the Lower Saxon Uplands or the Upper Weser area as being classified as landslide-prone. The local or regional exposure index estimates that 14.5 or 77.0 km of highways are at risk of landslides (Fig. 5.8). This means that on a local or regional level, 23.9 or 6.2 % of the highway network is located in potential landslide hazard area. The cost index calculated for the Upper Weser area amounts to US$52,000. By comparison, the cost figure, which only refers to total length of the local highway network, is about US$12,000. The average landslide costs for highways in the Lower Saxon Uplands are estimated at US$4.02 million per year.

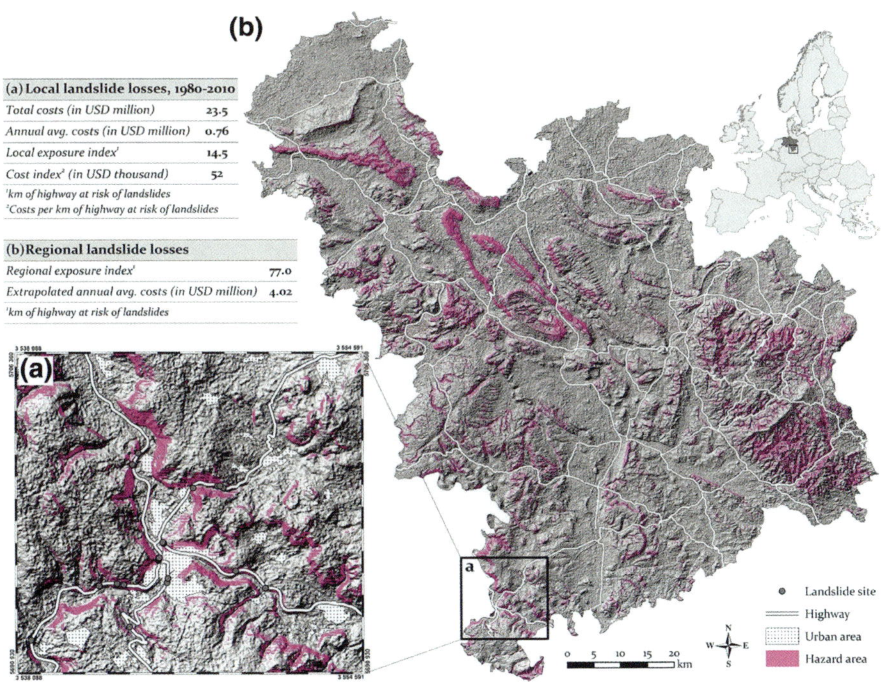

Fig. 5.8 Landslide losses for highways in the Upper Weser area and cost extrapolation for the entire highway network of the Lower Saxon Uplands (*Source* Klose et al., 2014b)

This corresponds to annual average costs per kilometer of highway of about US$3,200. A projection of the regional losses per year over the entire 31-year time period reaches the total of US$125 million.

5.3.2.2 Validation of the Cost Extrapolation

The plausibility of the cost extrapolation is tested by a validation of the landslide susceptibility model used to extrapolate landslide losses. This model validation is based on the concept of the success and prediction rate (Fig. 5.3; cf. Chung and Fabbri 2003). According to the success rate, 88 % of the landslides included in the model (modeling set = 85 % of the landslide inventory, 756 landslides) refer to the most susceptible 10 % of the total area. Alternatively, the prediction rate specifies that the most unstable 10 % of this region contains 92 % of the independent landslides (validation set = 15 % of the landslide inventory, 133 landslides). Although both rates prove good predictive power and high spatial accuracy, the results of the validation are no indicator of the overall plausibility of this model. This shows a conditional independence test between slope gradient and land use $\left(C_{corr} = 0.59; \chi^2 = 264.69 > \chi^2_{0.01}\right)$ and the fact that some lithological classes are not mutually exclusive (sand- and limestone, clay- and limestone). Despite such data related problems, which are difficult to avoid in studies with regional focus, the model is from a technical perspective suitable for proper spatial evaluation and zonation of landslide susceptibility for purposes of cost extrapolation.

5.3.3 Discussion

5.3.3.1 Methodological Problems and Solutions

The different tools of this methodological approach are designed and coordinated to meet crucial scale-related problems in ex post assessment of landslide losses for transportation infrastructures. Despite the fact that the test application of this methodology verifies its basic capacity for reliable cost estimation on local and regional levels, there are various methodological problems which need to be discussed in detail. Most of these problems relate to challenges of reduced data availability and quality when assessing landslide costs over broad areas and long time periods. The main problems of this methodology and ideas for its solution are presented in the following:

(a) Temporal cost volatility shows the importance of taking time periods of at least more than 10–20 years as a basis for reliable cost estimation in areas with low to moderate landslide activity. A major problem of cost modeling is that even comprehensive landslide database systems are characterized by a significant decrease in data completeness and quality over such long time periods (cf. Devoli et al. 2007; Hilker et al. 2009). This primarily affects the applicability of complex cost modules because of their comparatively high data requirements. Tools such as Google Earth© and Google Street View© are helpful to bridge some few data gaps (e.g., size of repair or mitigation structures), but yet, the landslide database for cost modeling needs to provide most of the necessary data without subsequent optimization. A solution to reduce data requirements in cost modeling is the concept of cost categories (Fig. 5.9). The idea is to replace LDMMs and cost modules by categories of the total costs for certain types of landslide damage. According to this concept, the costs of a landslide damage event are estimated by its classification to a specific cost category. Although this study provides losses for a number of typical landslide damage events at highways, the available data basis is yet too small to test or apply such preliminary cost categories in practice. A key advantage of cost modeling is to gain knowledge of major cost factors and drivers of landslide losses, which is vital for the development of reliable cost categories. As a result, the cost categories presented in Fig. 5.9 are likely to describe a reasonable range of costs, and this is most important for landslide cost assessment. Such a standardization of disaster management processes and landslide costs, however, is still associated with methodological problems. Thus, the application of both cost modeling and cost categorization is widely limited to less complex

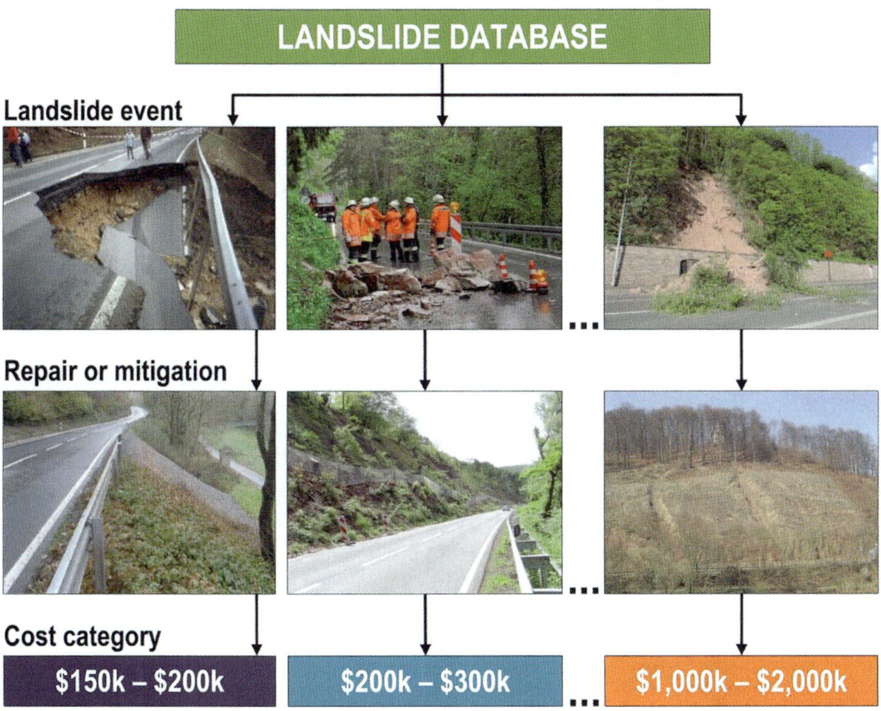

Fig. 5.9 Example of cost modeling based on categories of total costs for certain types of landslide damage (*Source* Klose et al., 2014b)

landslide damage events. By contrast, landslides causing exceptional damage to traffic routes (e.g., 2010 Taiwan Highway 3 Landslide; The Landslide Blog 2010) principally require individual assessment (i.e., cost survey), which conflicts with the idea of standardization that underlies both methodological approaches. Consequently, the use of this methodology is largely restricted to areas where transportation infrastructures are often affected by comparable types of landslide damage.

(b) A key component of this methodology is a regional landslide susceptibility model that enables to derive the main tools used for cost extrapolation. The development of regional landslide susceptibility models is generally faced with several methodological and data-related shortcomings (cf. Klose et al. 2014a). One problem of high relevance in this study is the limited availability of suitable input data for geofactors other than geomorphometry. As is the case with lithology and land use, the use of low-quality input data often implies further methodological problems, especially regarding the violation of conditional independence. A χ^2-test proves a significant spatial association between slope gradient and land use. This dependency, however, is tested to be of minor importance, as only leading to a constant overestimation of $SI(x)$, but not changing the statistical relations in the susceptibility scale. Consequently, none of these geofactors are excluded in landslide susceptibility modeling (cf. Neuhäuser et al. 2012a). The generalized geologic overview map (>1:500,000) is applied because medium to large-scale geologic maps are too precise in their class representation to support the calculation of useful statistical correlations. Nevertheless, this map shows the deficit of providing lithological classes that are not mutually exclusive, which conflicts with the assumption of conditional independence.

A further problem relates to the resolution of the ASTER GDEM of ± 25 m and its capacity to represent artificial slopes along traffic routes with sufficient spatial and topographic precision. Slope parameters derived from DEMs are negatively correlated to the size of their grid cells. This implicates a lack of exactness in relief representation, if applying a DEM with low to medium resolution (cf. Zhang et al. 1999). Several methodologies have been proposed to address this problem by data transformation and rescaling (cf. Qinke et al. 2008), but the slight systematic bias towards underestimating slope gradient (~5°–10°) identified in this study is widely negligible for cost extrapolation. Future research work has to address major weak points considering the concept of the exposure index. In its current usage, it simply defines an undifferentiated risk for transportation infrastructures located on a probable landslide mass but ignores hazard exposure in potential pathways of landslide movement. The integration of runout distances in regional landslide susceptibility models, however, exceeds the capabilities of today's modeling tools (cf. Klose et al. 2014a). Thus, the proposed exposure index will probably remain an incomplete concept that causes significant uncertainty in cost extrapolation.

(c) The result of the regional cost extrapolation is strongly influenced by the decision to operate with a cost index based on annual average costs. Due to outlier resistance, the annual median costs account for only US$17,000, which is about 45 times less the annual average cost of US$0.76 million. The use of annual median costs in cost extrapolation leads to a cost estimate for the Lower Saxon Uplands of less than US$0.1 million per year. This causes a difference in costs of almost US$4 million (98 %) compared to the cost extrapolation based on annual average costs. The large discrepancy between annual average costs and annual median costs is a consequence of comparatively high cost volatility and concentration, although on much shorter time scales than most other natural hazards. Thus, the annual return period of costs ≥ US$0.1 million is only $T = 3.1$ years on a local level, which affects short- to medium-term financial planning. This return period puts temporal cost volatility and concentration into perspective, but still maintains the need for analyzing annual cost trends over tens of years. A further source of error in cost estimation is related to the classification of insidious landslide damage or maintenance costs to one certain accounting year of the time series. As not distributing such costs over time, this study shows the tendency to slightly overestimate the volatility and concentration of annual landslide costs. Against this background, operating with annual average costs is proven to be a reasonable approach but requires keeping in mind high cost uncertainty.

5.3.3.2 Comparison of the Cost Estimate for the Lower Saxon Uplands

A comparison of the results of this cost assessment for highways in the Lower Saxon Uplands with landslide losses from study areas worldwide is strongly limited because of data scarcity. However, there are for some regions cost estimates available that support a comparison of costs, although such a comparison needs to be interpreted with caution. Some of the most recent data on annual landslide losses for highways are from the USA and include the states of Kentucky (>US$2 million; USGS 2013), Oregon (US$5.8 million; Wang et al. 2002), and Wyoming (US$1.0 million; Wyoming Homeland Security 2011). At the European level, Vranken et al. (2013) estimate for a regional case study area in the Ardennes, Belgium, annual average costs of landslide repair and/or mitigation for roads (US$0.8 million) and railways (US$0.6 million) of about US$1.4 million. The most comprehensive compilation of landslide losses for highways is still that of Walkinshaw (1992). On the basis of a cost survey for the US state highway systems (~20 % of the 1990 US highway network), this study provided an estimate of national landslide repair and maintenance costs of about US$190 million per year. Using the data sets published in this study, the costs per kilometer of highway are estimated at US$150, which strongly contrasts with the reference costs of US$3,200 for the Lower Saxon Uplands. The costs per kilometer of highway in the Lower Saxon Uplands are thus about 20 times as much as the US cost estimate. An additional study reports landslide repair costs for highways in the Lao PDR of, on average, US$7.2 million per year (cf. Hearn et al. 2008). According to the data presented in this study, the annual costs per kilometer of highway are about US$1,000. Although the difference in costs is less pronounced, the reference costs for the Lower Saxon Uplands are more than three times that of the Lao PDR. The comparison shows wide variations between the landslide losses of the different areas and proves that the cost estimates provided by this study are relatively high in value. However, the many influencing factors on these cost estimates (e.g., size of study area, level of landslide hazard, sociotechnical conditions) make their comparison difficult, which is why this comparison only supports a preliminary cross-checking of the obtained results.

References

Ackermann, E. (1953). Der aktive Bergrutsch südlich der Mackenröder Spitze in geologischer Sicht. *Nachrichten der Akademie der Wissenschaften in Göttingen, Mathematisch-Physikalische Klasse, IIa. Mathematisch-Physikalisch-Chemische Abteilung, 5*, 67–83.

Agterberg, F. P., & Cheng, Q. (2002). Conditional independence test for weights-of-evidence modeling. *Natural Resources Research, 11*(4), 249–255.

Ayalew, L., Yamagishi, H., & Ugawa, N. (2004). Landslide susceptibility mapping using GIS-based weighted linear combination, the case in Tsugawa area of Agano River, Niigata Prefecture, Japan. *Landslides, 1*, 73–81.

Backhaus, E., Gramann, F., Kaever, M., Lepper, J., Lohmann, H. H., Meiburg, P., et al. (1980). *Erläuterungen zur Geologischen Karte des Reinhardswaldes 1:50.000*. Wiesbaden: Hessisches Landesamt für Bodenforschung. 32 S.

Bălteanu, D., Chendeş, V., Sima, M., & Enciu, P. (2010). A country-wide spatial assessment of landslide susceptibility in Romania. *Geomorphology, 124*, 102–112.

Chung, C.-J. F., & Fabbri, A. G. (2003). Validation of spatial prediction models for landslide hazard mapping. *Natural Hazards, 30*, 451–472.

Conforti, M., Robustelli, G., Muto, F., & Critelli, S. (2012). Application and validation of bivariate GIS-based landslide susceptibility assessment for the Vitravo river catchment (Calabria, South Italy). *Natural Hazards, 61*, 127–141.

Crozier, M. J. (1999). Prediction of rainfall-triggered landslides: A test of the antecedent water status model. *Earth Surface Processes and Landforms, 24*, 825–833.

Dai, F. C., & Lee, C. F. (2002). Landslide characteristics and slope instability modeling using GIS, Lantau Island, Hong Kong. *Geomorphology, 42*, 213–228.

Damm, B. (2000). Hangrutschungen im Mittelgebirgsraum – Verdrängte "Naturgefahr"? *Standort, Zeitschrift für Angewandte Geographie, 24*, 27–34.

Damm, B. (2005). Gravitative Massenbewegungen in Südniedersachsen. Die Altmündener Wand – Analyse und Bewertung eines Rutschungsstandortes. *Zeitschrift für Geomorphologie NF, Suppl.-Bd. 138*, 189–209.

Damm, B., & Pflum, S. (2004). Geomorphologische Naturgefahren und Raumplanung – Bewertungsprobleme am Beispiel von Rutschgefahren in Südniedersachsen. *Zeitschrift für Geomorphologie NF, Suppl.-Bd. 135*, 127–146.

Damm, B., Varga, K., Heckmann, T., & Becht, M. (2009). The Impact of Bedrock Stratification on Landslide Susceptibility – An Example of GIS-based Landslide Modelling in the Bunter Sandstone Areas of Northern Hesse and Southern Lower Saxony (Germany). *Die Erde, 140*, 175–193.

Damm, B., Becht, M., Varga, K., & Heckmann, T. (2010). Relevance of tectonic and structural parameters in Triassic bedrock formations to landslide susceptibility in Quaternary hillslope sediments. *Quaternary International, 222*, 143–153.

Devoli, G., Strauch, W., Chavez, G., & Høeg, K. (2007). A landslide database for Nicaragua: A tool for landslide-hazard management. *Landslides, 4*, 163–176.

Fell, R., Corominas, J., Bonnard, C., Cascini, L., Leroi, E., Savage, W. Z., on behalf of the JTC-1 Joint Technical Committee on Landslides and Engineered Slopes. (2008). Guidelines for landslide susceptibility, hazard and risk zoning for land use planning. *Engineering Geology, 102*, 85–98.

Glade, T. (2000). Modelling landslide-triggering rainfalls in different regions of New Zealand – The soil water status model. *Zeitschrift für Geomorphologie NF, Suppl.-Bd. 122*, 63–84.

Godt, J. W., Baum, R. L., & Chleborad, A. F. (2006). Rainfall characteristics for shallow landsliding in Seattle, Washington, USA. *Earth Surface Processes and Landforms, 31*, 97–110.

Godt, J. W., Coe, J. A., Baum, R. L., Highland, L. M. (2012). Prototype landslide hazard map of the conterminous United States. In E. Eberhardt, C. Froese, K. Turner, & D. Leroueil (Eds.), *Landslides and engineered slopes: Protecting society through improved understanding* (pp. 245–250). London: Taylor & Francis Group.

Guillard, C., & Zezere, J. (2012). Landslide susceptibility assessment and validation in the framework of municipal planning in Portugal: The case of Loures Municipality. *Environmental Management, 50*, 721–735.

Guzzetti, F., Reichenbach, P., Ardizzone, F., Cardinali, M., & Galli, M. (2006). Estimating the quality of landslide susceptibility models. *Geomorphology, 81*, 166–184.

Havenith, H.-B., Strom, A., Caceres, F., & Pirard, E. (2006). Analysis of landslide susceptibility in the Suusamyr region, Tien Shan: Statistical and geotechnical approach. *Landslides, 3*, 39–50.

Hearn, G. J., Hunt, T., Aubert, J., & Howell, J. H. (2008). *Landslide impacts on the road network of Lao PDR and the feasibility of implementing a slope management programme*. International Conference on Management of Landslide Hazard in the Asia-Pacific Region, Sendai, Japan.

Hesemann, J. (1975). *Geologie Nordrhein-Westfalens*. Bochumer Geographische Arbeiten, Sonderreihe Band 2. Paderborn: Ferdinand Schöningh. 416 S.

Hilker, N., Badoux, A., & Hegg, C. (2009). The Swiss flood and landslide damage database 1972–2007. *Natural Hazards and Earth System Sciences, 9*, 913–925.

Jelínek, R., Maitan, S., & Omura, H. (2001). Slope movements in Slovakia – Geographic and geological characteristics. *Journal of the Faculty of Agriculture, Kyushu University, 45*(2), 589–600.

Klose, M., Damm, B., Terhorst, B., Schulz, N., & Gerold, G. (2012a). Wirtschaftliche Schäden durch gravitative Massenbewegungen. Entwicklung eines empirischen Berechnungsmodells mit regionaler Anwendung. *Interpraevent, 12*, 979–990.

Klose, M., Damm, B., & Gerold, G. (2012b). Analysis of landslide activity and soil moisture in hillslope sediments using a landslide database and a soil water balance model. *GEO-ÖKO, 33*(3–4), 204–231.

Klose, M., Gruber, D., Damm, B., & Gerold, G. (2014a). Spatial databases and GIS as tools for regional landslide susceptibility modeling. *Zeitschrift für Geomorphologie NF, 58*(1), 1–36.

Klose, M., Damm, B., & Terhorst, B. (2014b). Landslide cost modeling for transportation infrastructures: A methodological approach. *Landslides*. doi:10.1007/s10346-014-0481-1.

Komac, M. (2012). Regional landslide susceptibility model using the Monte Carlo approach – The case of Slovenia. *Geological Quarterly, 56*, 41–54.

Magliulo, P., Di Lisio, A., Russo, F., & Zelano, A. (2008). Geomorphology and landslide susceptibility assessment using GIS and bivariate statistics: A case study in Southern Italy. *Natural Hazards, 47*, 411–435.

Meyer, S. (2005). Der rezente Bergsturz am Messingsberg im niedersächsischen Wesergebirge. *Mitteilungen des Verbandes der Deutschen Höhlen- und Karstforscher, 51*(3), 96–100.

Nadim, F., Kjekstad, O., Peduzzi, P., Herold, C., & Jaedicke, C. (2006). Global landslide and avalanche hotspots. *Landslides, 3*, 159–173.

Neuhäuser, B., Terhorst, B., & Damm, B. (2012a). Landslide identification and modelling in Flysch areas of the European alpine foreland. *Zeitschrift für Geomorphologie NF, 56*(4), 115–146.

Neuhäuser, B., Damm, B., & Terhorst, B. (2012b). GIS-based assessment of landslide susceptibility on the base of the Weights-of-Evidence model. *Landslides, 9*, 511–528.

NNG. (2013). Bergrutsch Steinbergen. Norddeutsche Naturstein GmbH. Retrieved July 26, 2014, from http://www.bergrutsch-steinbergen.de.

Ohlmacher, G. C., & Davis, J. D. (2003). Using multiple logistic regression and GIS technology to predict landslide hazard in northeast Kansas, USA. *Engineering Geology, 69*, 331–343.

Pereira, S., Zêzere, J. L., Bateira, C. (2012). Technical note: Assessing predictive capacity and conditional independence of landslide predisposing factors for shallow landslide susceptibility models. *Natural Hazards and Earth System Sciences, 12*, 979–988.

Ponziani, F., Pandolfo, C., Stelluti, M., Berni, N., Brocca, L., & Moramarco, T. (2012). Assessment of rainfall thresholds and soil moisture modeling for operational hydrogeological risk prevention in the Umbria region (central Italy). *Landslides, 9*, 229–237.

Qinke, Y., Jupp, D., Rui, L., & Wie, L. (2008). Re-scaling lower resolution slope by histogram matching. In Q. Zhou, B. Lees, & G. Tang (Eds.), *Advances in digital terrain analysis* (pp. 193–210). Berlin: Springer.

Schunke, E. (1971). Die Massenverlagerungen an den Schichtstufen und Schichtkämmen des Weser-Leine-Berglandes. *Nachrichten der Akademie der Wissenschaften in Göttingen, II. Mathematisch-Physikalische Klasse, 3*, 1–35.

Schuster, R. L., & Highland, L. M. (2007). The third Hans Cloos lecture. Urban landslides: Socioeconomic impact and overview of mitigative strategies. *Bulletin of Engineering Geology and the Environment, 66*, 1–27.

Schwab, J. C., Gori, P. L., & Jeer, S. (Eds.). (2005). *Landslide hazards and planning* PAS Report 533/534. Washington D.C.: American Planning Association. 208 pp.

Sidle, R. C., & Ochiai, H. (2006). *Landslides: Processes, prediction, and land use*. Water Resources Monograph 18. Washington D.C.: American Geophysical Union. 312 pp.

The Landslide Blog. (2010). Images of the highway 3 landslide in Taiwan. In D. Petley (Ed.), *The Landslide Blog*. Retrieved December 10, 2013, from http://blogs.agu.org/landslideblog/2010/04/26/images-of-the-highway-3-landslide-in-taiwan/.

Thiery, Y., Malet, J.-P., Sterlacchini, S., Puissant, A., & Maquaire, O. (2007). Landslide susceptibility assessment by bivariate methods at large scales: Application to a complex mountainous environment. *Geomorphology, 92*, 38–59.

Tilch, N. (1999). Rutschungs-Suszeptibilität im südlichen Niedersachsen – Von der Anatomie der Rutschung bei Brunkensen/Alfeld zur Prognose instabiler Hanglagen. *Braunschweiger Geowissenschaftliche Arbeiten, 22*, 184 S.

Trigila, A., Iadanza, C. (2008). *Landslides in Italy* (32 pp. + Appendix). Special Report 2008. Rome: Institute for Environmental Protection and Research.

USGS. (2013). Landslide inventory pilot project – Kentucky. U.S. Geological Survey. Retrieved November 20, 2013, from http://landslides.usgs.gov/regional/inventory/kentucky/.

Van Beek, L. P. H., & Van Asch, Th. W. J. (2004). Regional assessment of the effects of land-use change on landslide hazard by means of physically based modelling. *Natural Hazards, 31*, 289–304.

Van Den Eeckhaut, M., Vanwalleghem, T., Poesen, J., Govers, G., Verstraeten, G., & Vandekerckhove, L. (2006). Prediction of landslide susceptibility using rare events logistic regression: A case-study in the Flemish Ardennes (Belgium). *Geomorphology, 76*, 392–410.

Van Den Eeckhaut, M., Hervás, J., Jaedicke, C., Malet, J.-P., Montanarella, L., & Nadim, F. (2012). Statistical modelling of Europe-wide landslide susceptibility using limited landslide inventory data. *Landslides, 9*, 357–369.

Van Westen, C. J. (2000). The modelling of landslide hazards using GIS. *Surveys in Geophysics, 21*, 241–255.

Van Westen, C. J., Castellanos, E., & Kuriakose, S. L. (2008). Spatial data for landslide susceptibility, hazard, and vulnerability assessment: An overview. *Engineering Geology, 102*, 112–131.

Varga, K., Becht, M., & Damm, B. (2006). Ansätze der GIS-gestützten räumlichen Modellierung von Rutschgefahren in Buntsandsteingebieten Nordhessens und Südniedersachsens (BRD). *Angewandte Geoinformatik, 2006*, 679–684.

Vranken, L., Van Turnhout, P., Van Den Eeckhaut, M., Vandekerckhove, L., & Poesen, J. (2013). Economic valuation of landslide damage in hilly regions: A case study from Flanders, Belgium. *Science of the Total Environment, 447*, 323–336.

Walkinshaw, J. (1992). Landslide correction costs on US state highway systems. *Transportation Research Record, 1343*, 36–41.

Wang, H. B., & Sassa, K. (2005). Comparative evaluation of landslide susceptibility in Minamata area, Japan. *Environmental Geology, 47*, 956–966.

Wang, Y., Summers, R. D., Hofmeister, R. J. (2002). *Landslide loss estimation pilot project in Oregon* (47 pp). Open-File Report O-02-05. Portland: Oregon Department of Geology and Mineral Industries.

Wyoming Homeland Security. (2011). Wyoming multi-hazard mitigation plan – Landslides. Retrieved November 20, 2013, from http://hls.wyogov/Library/mit_plan/11_Landslide.pdf.

Zhang, X., Drake, N. A., Wainwright, J., & Mulligan, M. (1999). Comparison of slope estimates from low resolution DEMs: Scaling issues and a fractal method for their solution. *Earth Surface Processes and Landforms, 24*, 763–779.

Chapter 6
Synthesis—Towards Integrated Assessment of Landslide Risk

6.1 Created Risk—The Role of Human Activity

Landslides in terms of their causes or triggers and the risk they entail on society show most often a natural geomorphic and human dimension (e.g., Sidle et al. 1985; Alexander 1989; Nadim et al. 2011; Sect. 1.1). These two dimensions are closely interconnected as a result of widespread urbanization and the global dominance of cultivated landscapes (e.g., Václavík et al. 2013; United Nations 2014). A large part of terrestrial landslides are thus occurring in a spatial setting where human activity is not just vulnerable to landslides but is also controlling their physical processes to some extent. Fundamental understanding of landslide risk requires therefore knowledge on how people are contributing to this risk by their own land use practices. The statistics and analyses on the causes of landslides and on landslide impact presented in this study revealed that the following aspects of landslide risk in Lower Saxony or the entire German Central Uplands make it necessary to turn towards an integrated risk assessment: (i) human activity is a major causative factor of landslides, not only by predisposing or triggering them, but also as a result of implementing inadequate or undersized mitigation measures; (ii) the level of tolerable or acceptable risk (see also Fell 1994), a measure driving a large part of landslide costs in Germany, is highly variable, differing between individuals or societies (Klose et al. 2014a), with its nature being to change over time; and (iii) decision makers often have difficulty in finding the right balance in hazard management (cf. Damm and Klose 2014), which implies lack of its effectiveness in both technical and financial terms, thus intensifying landslide risk in some cases. The proposed concept of integrated risk assessment takes into account these human-related or societal factors of landslide risk as they are critical for examining what defines landslide risk today and in future.

A key to assess landslide risk in the Lower Saxon Uplands in an integrated perspective is to track regional or site-specific infrastructure development over previous decades and to correlate it with past landslide activity (Fig. 6.1). The landslide database analyzed in this study indicates that different landslide sites throughout this region have been affected by landslides for more than the past 150 years.

M. Klose, *Landslide Databases as Tools for Integrated Assessment of Landslide Risk*, Springer Theses, DOI 10.1007/978-3-319-20403-1_6

Fig. 6.1 (continued)

20th Century
Further landslides and low-cost hazard mitigation of temporary effectiveness

About 30 landslide events occurred during the 20th century with clusters in 1924–1926, 1961, 1970, 1974, 1994, and 1999–2001 (e). Implementation of low-cost mitigation measures (catch fence, rockfall drapery, etc.) in response to these damage events, but hazard mitigation was usually undersized and failed under stress, wherefore not enabling effective risk reduction (f,g,h).

Fig. 6.1 (continued)

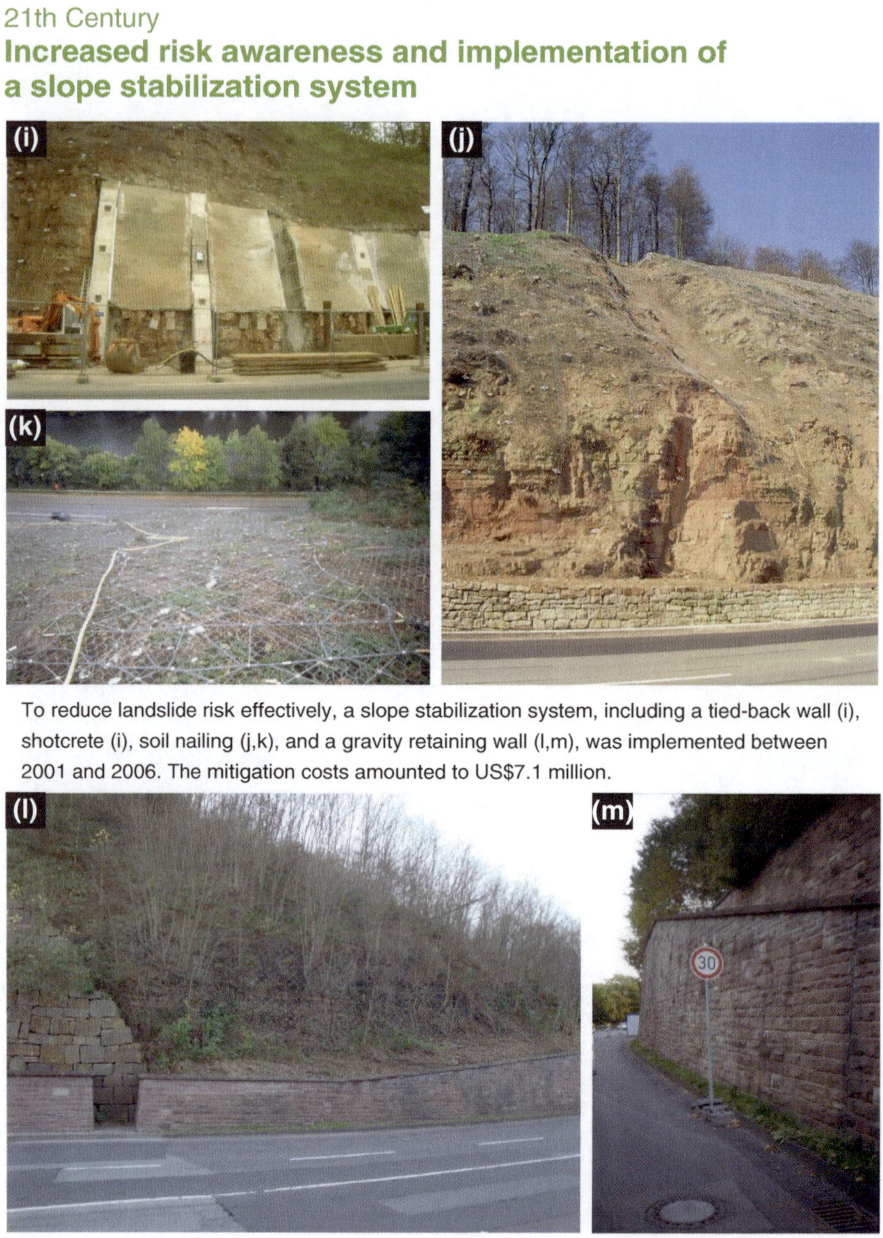

Fig. 6.1 Case history of landslide impacts for the landslide site "Altmündener Wand" at the highway B 3 in Hann. Münden, Upper Weser area (*Photos* Database B. Damm; M. Grochau, photo: g)

Landslide activity in the Lower Saxon Uplands generally intensified with increasing land use pressure since the mid-19th century (e.g., Wiese 1978; Von Pezold 1980). Population growth from then on resulted in an expansion of urban territory to areas with steep slopes that are naturally prone to landslides (cf. Damm 2005; Klose et al. 2014b). Hillside development in such areas often required massive slope disturbance by costly and sometimes inaccurate earthworks that frequently intensified the landslide predisposition of these areas or directly triggered landslides in many places. It was not until the early to mid-19th century when construction codes and methods considering the requirements for urban development and proper land use in unstable terrain became available (cf. Damm 2002, 2005). Lack of regulation or restriction of settlement as well as deficits in spatial planning in the past are often reasons for landslide damage to private homes and public infrastructure today. This applies especially to the landslide-prone city of Hann. Münden that is located in the Upper Weser case study area and that has already been part of many landslide studies (cf. Damm 2000, 2002, 2005, 2006; Damm and Pflum 2004; Klose et al. 2014b; Maurischat and Klose 2014).

The same holds true for the transportation system in this region that was characterized by rapid expansion since the mid-19th century (e.g., Baldermann 1968). Relief conditions in the Lower Saxon Uplands often require roads or railways to pass along the slopes of cuesta scarps or deep river valleys, which frequently made large cuts and fills necessary for building traffic routes (cf. Uhl 1907; Müller 1936). The embankments of many highways today are partly composed of non-draining fill materials from the early times of road construction in which it was common practice to build highway embankments using soil material from upslope or road excavation. These fill materials often originated from weak and moisture-sensitive Pleistocene slope deposits whose use is now prohibited by current construction codes (cf. Damm 2005; Floss 2006; Gidde 2012; see also Sect. 3.2). Together with cutting slopes at too steep an angle in areas with limited space for road location this has created a significant predisposition to landslides that continues to exist today and in future. Proper design and construction of highways as well as landslide hazard prevention are now widely established in regional transportation planning, especially with regard to new road development and renewal of existing highway infrastructure (see also Damm 2000, 2005). The analyses still have shown that most mitigation measures in recent years are implemented in response to periods of increased landslide activity instead of being realized proactively (cf. Maurischat and Klose 2014; Sect. 6.2). Despite an increased level of hazard awareness and adapted construction methods (Fig. 6.1), an effective protection of the many highway sections at risk of landslides has not been realized so far, as hazard mapping and field experience from this region indicates (cf. Damm 2005; Klose et al. 2014b, c).

An important role in integrated risk assessment is played by case histories of landslide impact that can be developed on the basis of information from landslide databases. Figure 6.1 presents an illustrated case history with focus on exemplifying the close dependency of landslide occurrence to human activity for a reference landslide site at a highway in the Upper Weser area. This case history of landslide impact provides profound insight into how people managed to live with landslide risk over centuries and how their land use decisions were affected by this risk or had contributed to intensifying it. The capability of such case histories for tracking the evolution of landslide hazard and risk in relation to human activity makes them a key for integrated risk assessment. Historical analyses like these, however, are limited to high-quality data sets, wherefore they are still exceptional and restricted to certain landslide sites. As a result, it is a basic requirement for databases used in this kind of landslide risk assessment to show at least at local level a widely complete landslide record, thus to rely on a top-down approach of data retrieval (cf. Sect. 2.2.2).

6.2 Are Landslides Economically Relevant?

Knowing how much landslides cost society is vital yet not sufficient for integrated assessment of landslide risk. The key question rather is whether calculated costs are economically relevant or not, which is equivalent to analyzing the fiscal relevance of the financial burden of landslide losses on public budgets. This specific field of risk assessment has been widely ignored in past research and is today in yet different form only addressed by Highland (2013) asking "who bears the burden" of landslide losses in the U.S. One major objective of the present work was therefore to model disaster financing and to assess the budgetary impact of landslide losses for highways and urban infrastructures in the Lower Saxon Uplands. Based on the results of the above cost estimation for highways and additional loss data published in Klose et al. (2012a), an economic impact study has been conducted using the example of the Lower Saxony Department of Transportation (NLStBV) and the city of Hann. Münden (~20,000 residents) in the Upper Weser area (Fig. 3.1). The methods used in this study included cost survey by interviewing urban planners and analysis of budget data at city level and from the NLStBV Regional Office Gandersheim.

Budgeting as part of fiscal planning in Lower Saxony involves, amongst others, two main steps (e.g., Rose 2008; Anders et al. 2008): (i) cost analysis, and (ii) risk identification. As showing the potential to affect public budgets, direct losses caused or expected to be caused by landslides should be taken into account in both steps, but this, however, is not the case in Lower Saxony, neither at municipal nor at state level. Financing of landslide damage is rather the opposite today, with hazard management being characterized by reactive instead of proactive thinking. The common practice can be best described by using a simple three-tier model which indicates that disaster financing is organized as ex-post financing based on budget reallocation (Fig. 6.2). A reserve of funds for unexpected expenditures related to landslide repair or mitigation generally does not exist in this model in which actions are taken and money is spent only if landslide damage results in disruption of infrastructure. Disaster response as first step of the disaster cycle is financed with cash from operating budgets of which maintenance depots

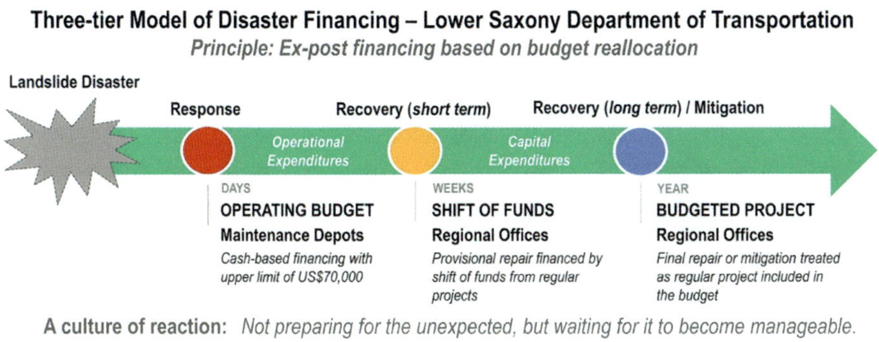

Fig. 6.2 The standard procedure of landslide disaster financing in the public infrastructure sector in Lower Saxony. Using the example of the Lower Saxony Department of Transportation (NLStBV), the flowchart describes the main steps of disaster financing with the affected budgets and the types of expenditures (*Source* according to Maurischat and Klose 2014)

are responsible for. The upper limit of cash-based disaster financing in transportation planning is set to a threshold value of around US$70,000. In case of landslide damage exceeding this cost limit, payments correspond to capital expenditures, with the NLStBV regional offices having the responsibility for any kind of investment. To avoid long-term disruption of both urban and highway infrastructure, provisional repair or mitigation is the starting point of subsequent disaster recovery. The only way to finance such immediate measures is to shift funds from regular projects that are already considered in budgeting. Temporary solutions are replaced by final ones when enough time has passed to treat repair or mitigation as regular projects whose costs are included in annual budgets. This last step of disaster financing implicates that repair or mitigation is paid in the same way as ordinary construction works. A key principle of hazard management defining the prevailing culture of reaction is thus not to prepare for losses in future but waiting for incurred losses to become manageable (see also Maurischat and Klose 2014).

The results of the impact study indicate that landslide losses affect public budgets in temporally variable intensity (Fig. 6.3). Much of the reference costs relate to insidious landslide losses or maintenance expenditures that have been classified to certain accounting years of which some are in future. As not distributing such costs over time, there is the tendency to overestimate cost volatility, which is important to keep in mind when interpreting Fig. 6.3. A cost survey for the city of Hann. Münden has shown that landslide losses at city level amount up to US$2.4 million (urban roads) or US$1.0 million (sewer systems) in costly years (Fig. 6.3b, c). These annual cost extremes correspond to 141 % (urban roads) or 47 % (sewer systems) of the relevant annual budget for regular projects. The average financial burden of landslides in relation to such annual reference budgets accounts for 44 % (urban roads) or 20 % (sewer systems) between 2010 and 2015. Expert interviews revealed that slope creep reduces the average life cycle of sewer lines at about 50 %, and the costs of earthworks in this city are

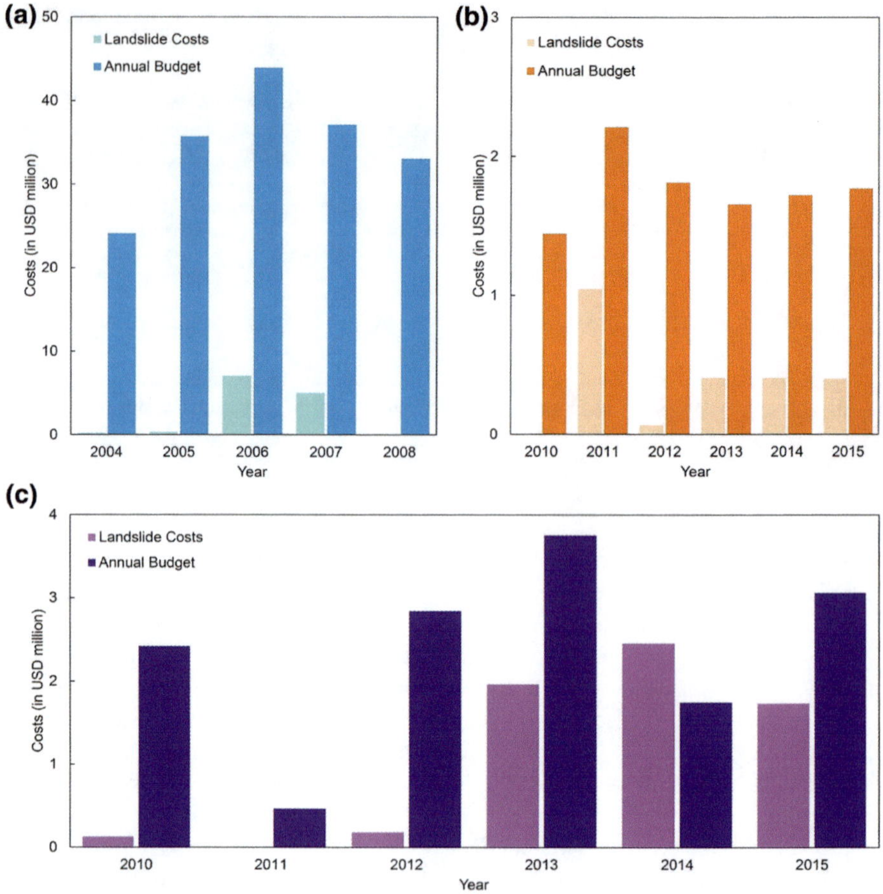

Fig. 6.3 Annual comparison of direct landslide costs and regular construction budgets for different types of public infrastructure in the Lower Saxon Uplands. The presented data serve as an indicator for the temporarily high financial burden of landslide losses and refer to **a** the highway network of the Regional Office Gandersheim, Lower Saxony Department of Transportation (NLStBV), and **b** sewer systems or **c** urban roads in the city of Hann. Münden, Upper Weser area. Note that the losses shown in (**a**), in contrast to the budget data, relate only to the Upper Weser area that covers about 20 % of the highway network operated by the NLStBV Regional Office Gandersheim. The losses for 2014 and 2015 shown in (**b**) and (**c**) refer to previous landslide events whose repair will take place in 2014 and 2015. These damage costs are therefore already included in the future construction budgets (*Source* according to Maurischat and Klose 2014)

increased by a factor of 1.4 on average. City planners estimate the annual maintenance costs for urban roads in Hann. Münden to be at US$1.5 million, which is probably about two to three times the average costs of cities with comparable size. The annual average direct costs for highways in the Upper Weser area, which covers about 20 % of the road network of the NLStBV Regional Office Gandersheim, amount to US$0.76 million between 1980 and 2010. These costs correspond to 2.2 % of the annual average construction budget (US$34.8 million) in the period 2004–2008 (Fig. 6.3a). In costly years like 2006 (US$7.1 million) or 2007 (US$5.1 million), however, the regional office was required to pay costs for landslide repair or mitigation equivalent to up to 16 % of the regular construction budgets (cf. Maurischat and Klose 2014).

The two case studies clearly illustrate that landslides show the potential to burden public budgets significantly, constituting a substantial cost factor in short- to mid-term fiscal planning. Given these research results, an economically justified argument for either reactive or proactive hazard management is yet hard to find by now, but one thing seems to be certain: a culture of reaction contradicts the basic principles of fiscal planning and is very likely accompanied by significant opportunity costs.

6.3 Conclusions

This study applies the landslide database for the Federal Republic of Germany in different fields of regional landslide risk assessment. Using the example of the Lower Saxon Uplands, the research work presents approaches and methods for analysis of spatial and temporal landslide hazard in this mountain area. A regional landslide susceptibility model has been developed to estimate the spatial probability of landslides, whereas assessment of temporal landslide probability was realized by optimization and application of a soil water balance model for regional simulation of landslide-causative soil moisture levels. The study also deals with an estimation of direct landslide costs for highways in the Lower Saxon Uplands. This part of the research work involved the design and application of a new method for landslide cost modeling with various tools to compile, model, and extrapolate landslide losses on different spatial scales over time. A further part of the present study investigates the role of human activity in landslide hazard or risk and has a special focus on analyzing the financial burdens of landslides on public budgets. The research in this regard combines modeling of landslide disaster financing with a budgetary impact study based on cost survey and expert interviews. All these different parts of this study support drawing the following final conclusions on landslide hazard and risk in the Lower Saxon Uplands:

Landslides in the study area are spatially clustered according to the local and regional relief, with a significant level of landslide hazard at locations featuring steep terrain, including cuesta scarps, V-shaped or gorge-like valleys, and river cut banks. The spatial distribution of hazard area is also characterized by isolated clusters, some of which corresponding with single hills, scarps, terraces, and large artificial slopes (see also Damm 2005; Gruber 2012). Most landslides have been identified in areas with Mesozoic sedimentary rock, especially at sites where permeable lime- or sandstones are on top of weak marl- and claystones (cf. Klose et al. 2014b, c). A large number of landslides in this region are connected with prolonged wet periods of which ones that result in landslide-causative soil moisture levels show recurrence intervals of 0.9 years. During such wet periods the probability of slope failure ranges between 10 and 47 %, wherefore landslides are likely to occur on an annual basis in above hazard areas (cf. Klose et al. 2012b). This applies especially to the Upper Weser area where moisture-sensitive and landslide-prone periglacial sediments are widespread on many steep slopes (see also Damm 2005). Urban expansion and associated land use pressure over the past century resulted in the development of many hillsides at risk of landslides, implicating that urban area and hazard zones often spatially coincide today. Local communities throughout this region are thus exposed to landslides in many places, whereby some of these vulnerable towns or cities are located with 10–20 % of their area in zones of significant hazard. Alternatively, the regional highway network crosses areas susceptible to landslides on a length of 77 km, which corresponds to 6 % of its total length in the Lower Saxon Uplands (cf. Klose et al. 2014b, d). Landslides in this mountain area are closely connected with human activity, not only because of urban expansion into hazard areas or improper land use practices, but also because of inadequate risk management, especially the use of low-cost mitigation measures that are often undersized, thus tend to fail under stress (see also Damm and Klose 2014, 2015). These deficits in past transportation planning are partly responsible for direct costs of landslide damage to highways in the amount of US$4.02 million per year. In the Lower Saxon Uplands, landslides are frequent yet often less costly damage events, with more than 80 % of them that affect highways resulting in costs lower than US$0.1 million. These minimum reference costs for highways in this region show a return period of 3.1 years and are at least at local level a significant cost factor in short- to mid-term fiscal planning (cf. Klose et al. 2014d). The financial burdens of landslide losses for urban or highway infrastructures at local or regional level vary between 2 and 44 % of exemplary annual average construction budgets and are thus of high economic relevance (see also Maurischat and Klose 2014).

In conclusion, landslide risk in economic terms is evaluated to be significant in the Lower Saxon Uplands, or more precisely, to be of low to medium level when using expert opinion to classify risk in a qualitative way. This personal judgement is not the result of systematic risk measurement on a numerical scale but refers to a qualitative assessment based on the research results presented in the previous sections. Even from a global perspective (cf. Sect. 1.2), these findings contrast with the outcomes of previous continental-scale landslide hazard and risk assessments, classifying the Lower Saxon Uplands or related Central European low mountain areas to show negligible or very low levels of landslide hazard and risk (e.g., Dilley et al. 2005; Nadim et al. 2006; Van Den Eeckhaut et al. 2012). Given the data and results presented in this study, the main reasons for not classifying the Lower Saxon Uplands as high-risk area are as follows: (i) dominance of small to moderate landslide magnitudes, (ii) low risk of loss of life due to landslides, and (iii) high technological and financial coping capacity for landslide risk reduction. This classification considering the extent and severity of landslide impact worldwide yet requires to be interpreted with caution. Landslides in the study area are widespread everyday hazards and as a whole have a considerable societal impact at local and regional level. The research work has revealed that landslides in "low-risk" areas such as the Lower Saxon Uplands are a critical cost factor in fiscal planning that shows the potential to burden public budgets substantially. These findings of the conducted study thus close an important knowledge gap and justify the urgent need for further research on landslide hazard and risk by demonstrating the high economic relevance of landslides.

The above conclusion represents an expert-based risk assessment that rests upon quantitative analyses and statistics on both natural and human factors of landslide risk. Despite analyzing spatiotemporal landslide hazard and economic landslide impact in a quantitative way, the final risk estimate provided by this assessment is still qualitative. The problem that continues to exist mainly relates to finding a systematic approach for integrated risk assessment that enables to combine the results of the different studies and to integrate them in an objective, transparent, and reproducible risk estimate. A major challenge thereby is to consider the results of historical analyses and soft risk factors such as hazard awareness and risk acceptance. These qualitative dimensions of landslide risk play a vital role in defining the overall level of risk but are beyond functional expression in a risk equation so far. In transportation planning, for example, low risk aversion in the present is equivalent to reduced mitigation efforts, with the result that near-future prevention cost will be at minimum level, whereas damage and disruption cost could reach a maximum in the long run. To take account of such relationship in risk assessment, there is the need for a system that contains rules to weight such qualitative indicators and to make them compatible with quantitative hazard and loss data. Further research work is thus necessary to systematically obtain a risk estimate that also integrates these qualitative findings, thus provides a rigorous measure of landslide risk on the basis of how people have dealt with landslides in the past.

A further key question of future research is how to optimize the proposed framework of integrated risk assessment to enable the calculation of a probabilistic monetary estimate of landslide risk. A promising approach combining an ex-post assessment with prediction of future losses is the Probabilistic Landslide Assessment Cost Estimation System (PLACES), a statistical modeling tool for probabilistic cost calculation developed by Crovelli and Coe (2008, 2009). The idea of PLACES is to predict the number of future landslides and their direct losses based on recurrence intervals and average costs of past landslides. While PLACES provides functionality for cost projection into the future, the approach presented in this study provides tools to bridge scale differences in cost assessment at large spatial scales. Future methodological research should thus take advantage of both approaches that in combination are an important step towards integrated risk assessment over broad areas. A fusion of these two toolsets for cost analysis, however, still does not solve a main obstacle in the application of such integrated approaches, which is their requirement of high-quality landslide databases.

The studies presented herein have shown that landslide databases are valuable sources of information for research on landslides, not only in terms of their causes, types, and processes, but also the impacts and costs associated with them. In order to serve as useful tools for integrated assessment of landslide risk, the development of landslide databases requires relying on a targeted strategy of data retrieval that in form of a bottom-up approach condenses data coverage from national or regional to local level. Together with systematic extraction and integration of data from multiple sources, this allows the development of landslide databases with a broad potential of application in integrated risk assessment. In-depth analysis of such landslide databases using the methods presented in this study enables to open a whole new window on landslide risk and is thus a key to promote effective disaster risk reduction in landslide practice.

References

Alexander, D. (1989). Urban landslides: *Progress in Physical Geography, 13*(2), 157–189.

Anders, R., Horstmann, J., Bernhardt, H., Mutschler, K., Stockel-Veltmann, C. (2008). *Kommunales Finanzmanagement in Niedersachsen – Neues Kommunales Rechnungswesen (NKR). Fachbuch mit praktischen Übungen und Lösungen.* Verlag Bernhardt-Witten, Witten, 717 S.

Baldermann, U. (1968). *Die Entwicklung des Straßennetzes in Niedersachsen von 1768–1960.* Reihe A: Forschungen zur Landes- und Volkskunde 87. Niedersächsisches Institut für Landeskunde und Landesentwicklung, Universität Göttingen. 114 S.

Crovelli, R. A., Coe, J. A., 2008. *Probabilistic methodology for estimation of number and economic loss (cost) of future landslides in the San Francisco Bay region, California* (27 pp). Open-File Report 2012–1204. Reston: U.S. Geological Survey.

Crovelli, R. A., & Coe, J. A. (2009). Probabilistic estimation of numbers and costs of future landslides in the San Francisco Bay region. *Georisk, 3*, 206–223.

Damm, B. (2000). Hangrutschungen im Mittelgebirgsraum – Verdrängte „Naturgefahr"? *Standort, Zeitschrift für Angewandte Geographie, 24*, 27–34.

Damm, B. (2002). Rutschungen im Fulda- und Oberweserraum (Nordhessen/Südniedersac hsen) – Ursachen, Auslöser und zeitliche Häufungen. In F. Fiedler, F. Nestmann, & M. Kohler (Eds.), *Naturkatastrophen in Mittelgebirgsregionen* (pp. 129–147). Berlin: Verlag für Wissenschaft und Forschung.

Damm, B. (2005). *Gravitative Massenbewegungen in Mittelgebirgsräumen – Geowissenschaftlich-historische Analyse von Massenbewegungsprozessen in Nordhessen und Südniedersachsen als Beitrag zur Angewandten Geomorphologie* (262 S. + Anhang, unveröffentlicht). Habilitationsschrift, Universität Eichstätt-Ingolstadt.

Damm, B. (2006). Simplified planning regulations in Germany and the effects of landslides. *Journal of Environmental Planning and Management, 49*, 777–790.

Damm, B., & Pflum, S. (2004). Geomorphologische Naturgefahren und Raumplanung – Bewertungsprobleme am Beispiel von Rutschgefahren in Südniedersachsen. Zeitschrift für Geomorphologie NF, *Suppl.-Bd. 135*, 127–146.

Damm, B., & Klose, M. (2014). Landslide database for the Federal Republic of Germany: A tool for analysis of mass movement processes and impacts. In K. Sassa, P. Canuti, & Y. Yin (Eds.), *Landslide Science for a Safer Geoenvironment, Vol. 2: Methods of Landslide Studies* (pp. 787–792). Berlin: Springer.

Damm, B., & Klose, M. (2015). The landslide database for Germany: Closing the gap at national level. *Geomorphology*. doi:10.1016/j.geomorph.2015.03.021.

Dilley, M., Chen, R. S., Deichmann, U., Lerner-Lam, A. L., Arnold, M., Agwe, J., et al. (2005). *Natural disaster hotspots: A global risk analysis*. Washington D.C.: The World Bank. 132 pp.

Fell, R. (1994). Landslide risk assessment and acceptable risk. *Canadian Geotechnical Journal, 31*(2), 261–272.

Floss, R. (2006). *Zusätzliche Technische Vertragsbedingungen und Richtlinien für Erdarbeiten im Straßenbau: ZTVE-StB 94, Fassung 1997. Kommentar mit Kompendium Erd- und Felsbau unter Berücksichtigung der nationalen und europäischen Regelwerke.* Bonn: Kirschbaum. 660 S.

Gidde, A. (2012). Niederschlagsabhängige Böschungsschäden in Niedersachsen (S. 90–92). In: Bundesanstalt für Straßenwesen (Hrsg.), *42. Erfahrungsaustausch über Erdarbeiten im Straßenbau.* Berichte der Bundesanstalt für Straßenwesen, Heft S 76. Bremerhaven: Wirtschaftsverlag NW.

Gruber, D. (2012). *GIS-gestützte Identifikation und Analyse der räumlichen Verbreitung gravitativer Massenbewegungen in Niedersachsen* (72 S. + Anhang). Masterarbeit, Georg-August-Universität Göttingen, Geographisches Institut.

Highland, L. M. (2013). *Direct landslide costs in the United States: Who bears the burden? A pilot landslide loss study for the states of Washington and Oregon with an application to the 50 United States.* GSA Annual Meeting 2013, Denver (CO), USA. Paper No. 106-1. Retrieved August 6, 2014, from https://gsa.confex.com/gsa/2013AM/webprogram/Paper228608.html.

Klose, M., Damm, B., Terhorst, B., Schulz, N., & Gerold, G. (2012a). Wirtschaftliche Schäden durch gravitative Massenbewegungen. Entwicklung eines empirischen Berechnungsmodells mit regionaler Anwendung. *Interpraevent, 12*, 979–990.

Klose, M., Damm, B., & Gerold, G. (2012b). Analysis of landslide activity and soil moisture in hillslope sediments using a landslide database and a soil water balance model. *GEO-ÖKO, 33*(3–4), 204–231.

Klose, M., Highland, L., Damm, B., & Terhorst, B. (2014a). Estimation of direct landslide costs in industrialized countries: Challenges, concepts, and case study. In K. Sassa, P. Canuti, & Y. Yin (Eds.), *Landslide Science for a Safer Geoenvironment, Vol. 2: Methods of Landslide Studies* (pp. 661–667). Berlin: Springer.

Klose, M., Gruber, D., Damm, B., & Gerold, G. (2014b). Spatial databases and GIS as tools for regional landslide susceptibility modeling. *Zeitschrift für Geomorphologie NF, 58*(1), 1–36.

Klose, M., Gruber, D., Damm, B., & Gerold, G. (2014c). Landslide susceptibility modeling on regional scales: The case of Lower Saxony, NW Germany. In K. Sassa, P. Canuti, & Y. Yin (Eds.), *Landslide Science for a Safer Geoenvironment, Vol. 2: Methods of Landslide Studies* (pp. 437–442). Berlin: Springer.

Klose, M., Damm, B., & Terhorst, B. (2014d). Landslide cost modeling for transportation infrastructures: A methodological approach. *Landslides.* doi:10.1007/s10346-014-0481-1.

Maurischat, P., Klose, M. (2014). Financing and budgetary impact of landslide losses for highways and urban infrastructures in NW Germany – An economic analysis using landslide database information and cost survey data. *Geophysical Research Abstracts*, 16, EGU2014-1995.

Müller, H. (1936). *Die Eisenbahnen im Gebiet der Oberweser.* Reihe A, Heft 32. Wirtschaftswissenschaftliche Gesellschaft zum Studium Niedersachsens. 89 S.

Nadim, F., Kjekstad, O., Peduzzi, P., Herold, C., & Jaedicke, C. (2006). Global landslide and avalanche hotspots. *Landslides, 3*, 159–173.

Nadim, F., Høydal, Ø., Haugland, H., McLean, A. (2011). *Analysis of landslides triggered by anthropogenic factors in Europe* (82 pp). SafeLand FP7, Deliverable D1.6. Retrieved August 13, 2014, from http://www.safeland-fp7.eu/results/Documents/D1.6.pdf.

Rose, J. (2008). *Kommunale Finanzwirtschaft Niedersachsen: Grundriss für die Aus- und Fortbildung.* Kiel: Kohlhammer. 568 S.

Sidle, R. C., Pearce, A.. J., O'Loughlin, C. L. (1985). *Hillslope stability and land use.* Water Resources Monograph 11. Washington D.C.: American Geophysical Union. 140 pp.

Uhl, B. (1907). Die Verkehrswege und Flußtäler um Münden und ihr Einfluß auf Anlage und Entwicklung der Siedlungen. *Forschungen zur Geschichte Niedersachsens, I. Band, 4. Heft*, 52 S.

United Nations (2014). *World Urbanization Prospects: The 2014 Revision, Highlights.* United Nations, Department of Economic and Social Affairs. 19 pp. + Appendix.

Václavík, T., Lautenbach, S., Kuemmerle, T., & Seppelt, R. (2013). Mapping global land system archetypes. *Global Environmental Change, 23*, 1637–1647.

Van Den Eeckhaut, M., Hervás, J., Jaedicke, C., Malet, J.-P., Montanarella, L., & Nadim, F. (2012). Statistical modelling of Europe-wide landslide susceptibility using limited landslide inventory data. *Landslides, 9*, 357–369.

Von Pezold, J. D. (1980). *Münden im 19. Jahrhundert: Verkehrsverhältnisse – äußeres Erscheinungsbild – Bevölkerung.* Stadt Münden. 63 S.

Wiese, H. (1978). Industrie und Stadtentwicklung ausgewählter Kleinstädte Südniedersachsens: Alfeld, Einbeck und Northeim. *Forschungen zur niedersächsischen Landeskunde, Band 110*, 111 S.

Printed by Printforce, the Netherlands